DARWIN STRIKES BACK

Defending the Science of Intelligent Design

THOMAS WOODWARD

BakerBooks
Grand Rapids, Michigan

Published by Baker Books
a division of Baker Publishing Group
P.O. Box 6287, Grand Rapids, MI 49516-6287
www.bakerbooks.com

Printed in the United States of America

Library of Congress Cataloging-in-Publication Data
Woodward, Thomas, 1950–
Darwin strikes back : defending the science of intelligent design / Thomas Woodward.
p. cm.
Includes bibliographical references.
ISBN 10: 0-8010-6563-1 (pbk.)
ISBN 978-0-8010-6563-7 (pbk.)
1. Intelligent design (Teleology) 2. Creationism. 3. Evolution (Biology) 4. Evolution (Biology)—Religious aspects—Christianity. 5. Religion and science. I. Title.
BL263.W67 2006
576.8′2—dc22 2006025001

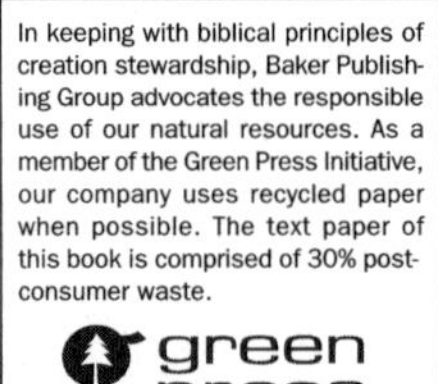

"*Darwin Strikes Back* tells the thrilling story of how the Darwinian establishment has summoned all its power to crush the frightening challenge of the Intelligent Design Movement, and how the rebels are not only surviving but gaining new strength as we respond to the onslaught. Highly recommended."

Phillip E. Johnson, emeritus professor of law, University of California, Berkeley; author, *Darwin on Trial*

"Taking the reader behind the headlines, Thomas Woodward—the premier historian of the Intelligent Design Movement—analyzes crucial developments of the past decade."

Michael J. Behe, department of biology, Lehigh University

"The controversy over Darwinism and Intelligent Design signals a major scientific and social revolution. Everyone who wants to understand it should read this timely and well-written book."

Jonathan Wells, author, *The Politically Incorrect Guide to Darwinism and Intelligent Design*

"My friend Tom Woodward is one of the most effective and articulate chroniclers of Intelligent Design in America today. This excellent work will clear away some of the fog of confusion spread by Darwinists, and give you solid, useable information to defend Intelligent Design. A very valuable resource in one of the most consequential debates of our times."

Chuck Colson, founder and chairman, Prison Fellowship

"Lucid, thorough, and brisk as the morning news, *Darwin Strikes Back* traces the launch of the Intelligent Design Movement and the response it has elicited. Woodward shows how ID challenges the interpretation of intelligent display without intelligent agency and calls for an alternative assessment of scientific data. While providing a valuable resource for the seasoned observer, this book should be especially appealing to students and newcomers to the debate wishing to be rapidly brought up to speed."

Leo R. Zacharski, professor of medicine, Dartmouth Medical School

"In their attempt to return fire in the debate against Intelligent Design, the Darwinists are mostly shooting blanks. Tom Woodward does a masterful job of dissecting weak polemic and showing how the Darwinian establishment has resorted to proof by confident assertion, genetic fallacies, and ad hominem tactics rather than genuinely engaging the arguments and evidence that ID theorists have mustered. Woodward predicts correctly that such tactics will not ultimately prevail."

Stephen C. Meyer, director, Center for Science and Culture Discovery Institute

"In *Darwin Strikes Back* Tom Woodward chronicles the recent acrimonious history of ID and its antagonists. Woodward is an insider who tells an engaging story that will clarify both the nature and the source of current sharp debate surrounding this issue."

Kenneth Petzinger, professor of physics, College of William and Mary

"In *Darwin Strikes Back*, Woodward presents a clear, accurate, and intriguing account of Intelligent Design, its history, the arguments in its favor, the counterarguments by the Darwinists, and the responses by the ID theorists. This is an important book for anyone who wants a clear picture of the ID/evolution debate."

Russell W. Carlson, professor of biochemistry and molecular biology; executive technical director of the complex carbohydrate research center, University of Georgia

"This is an important book. It brings us up to date on the latest round of skirmishing in the ever-continuing debate over our origins, and helps set the stage for the next round. Better yet, it makes clear to those who may not have followed recent events that the real scientific debate began well but was cut short early, and for the most part still remains unresolved."

David Keller, professor of chemistry, University of New Mexico

"Dr. Woodward has done a favor to both the sympathizers and detractors of the ID movement. Although Tom is clearly an ID proponent, he retains the ability to step back and allow the anti-ID critics to present their case. This is quickly followed by the rebuttals of the leading ID figures. He covers all the leading criticisms: 'ID is not science,' 'ID is religious,' 'Irreducible complexity has been refuted,' and even the more imaginative 'ID will be the end of science.' No one has a better grasp of the ID movement and its detractors than Tom Woodward."

Ralph Seelke, professor of biology and earth science, University of Wisconsin-Superior

"A brilliant and exhaustive sequel to *Doubts about Darwin*, critiquing the challenges to ID from 1996 to 2006 and documenting a pattern in these responses that is surprisingly long on rhetoric and short on science, by one of the leading proponents of Intelligent Design."

Walter Bradley, distinguished professor of mechanical engineering, Baylor University

"In *Darwin Strikes Back*, Tom Woodward has done the ID community an invaluable service. The progress of the ID Movement can be monitored most clearly by the increasing attacks from the evolutionary community. Those outside the battle may be tempted to think ID has been tripped up and is in retreat. Woodward's lucid prose and detailed research shows just the opposite. Read and be encouraged!"

Raymond G. Bohlin, lecturer in evolution; president, Probe Ministries

Contents

Foreword

Like a spy in a John Le Carré novel who has attended every crucial event in the Cold War, Tom Woodward has been ubiquitous in the unfolding culture war over intelligent design. He is the insiders' insider. With *Doubts about Darwin*, he established himself as the historian of the Intelligent Design Movement. Now, with *Darwin Strikes Back*, he also assumes the role of a gifted war correspondent, moving up and down the lines of engagement, tracing streams of intense and often ferocious rhetoric as they are poured out upon design theorists by panic-stricken Darwinists.

I first met Tom in 1990 while I was a postdoctoral fellow in computer science at Princeton University. As an alumnus of Princeton, he began working closely with a group of Princeton faculty members in 1988, with the goal of developing an annual lecture series at the university on a variety of academic topics. Together with these professors, he organized lectures by Alvin Plantinga at Princeton in the fall of 1990. I met Tom at one of these lectures, and in the coming years we experienced such a "university campus rendezvous" in many other places, especially as he played a key role in bringing Darwinian scholars and design theorists together in frank exchange and mutual critique.

It is fitting that we met at a lecture by Alvin Plantinga, since Plantinga is not just one of the most highly regarded philosophers of our era; he is also one who has written sympathetically about the intellectual project of Intelligent Design. In this context, he can be viewed as a symbol of the spiraling rhetorical nightmare faced by neo-Darwinism in the high university world. The nightmare is not simply the result of political pressure that Darwinists are experiencing. Rather, it is that the Darwinian account of evolution on which they are pinning their hopes is imploding.

Arguments for design are based on empirically identifiable patterns in the universe and demonstrate that intelligence is an essential aspect of the known causal structure of the universe (see the "can do premise" in the chapters that follow). In consequence, design inferences cannot be easily dismissed with furious bluster, or an ad hominem "wave of the hand," or even theological invocations of "poor design." In fact, as Woodward points out in this volume, the counter-rhetoric of Darwin's defenders is lurching into a mode so strident and vitriolic as to provoke more curiosity about the psychological causes of Darwinists' emotional states than about the "evil motives" of ID advocates. Historians of science regularly help us to understand this sort of personal subtext of scientific argument, but as a rhetorical historian, Woodward has done even more: he has explored this side of the debate with special care, cataloguing with vivid and unforgettable detail the labyrinth where logic and empirical evidence meet emotion and personal narrative.

Woodward's previous work—*Doubts about Darwin*—received the high regard of numerous scholars not at all associated with the ID Movement (see the "unexpected allies" in chapter 11 of this book). Likewise, in *Darwin Strikes Back*, his narratives and insights as a rhetorician of science should prove just as indispensable for the defenders of Darwinian orthodoxy as they are for the challengers. I predict this will especially be the case in his analysis of the debate swirling around Michael Behe and the flagellum (chapter 5), as well as his coverage of the origin of life stalemate (chapters 8 and 9) and the self-maiming explosives of the atheologians (chapter 11).

It's been said that cultural and intellectual movements go through three stages: first, they are ridiculed; second, they are violently opposed; and third, they are accepted as second nature so that people can't even imagine what the fuss was all about. In this book, Woodward shows how the ID movement has now entered the second stage, and then he assesses how we are doing. Stage two is the critical stage. It's at this stage that the future of a movement is decided—whether it has what it takes to weather the withering criticisms that are brought against it or whether it will bite the dust.

Woodward is optimistic, as am I, about the ultimate outcome of the controversy over ID, and he concludes his careful analysis with some pretty daring predictions. If he is right, we may look forward to a third volume from his hand, one to complete a trilogy on the ID movement that started with *Doubts about Darwin* and now has issued in *Darwin Strikes Back*. If he is right, this third volume could appropriately be called *The Triumph of Design*. But since "Darwin" figures in the titles of previous volumes in the trilogy, he may want to go with something like *Darwin's Doddering Idea* or *Darwinism—The Senescent Years*.

William A. Dembski

Preface

The Big Picture

On the first of August 2005, reporters at the White House asked President Bush his opinion about the new Intelligent Design theory that had appeared with increasing frequency in the news. Specifically: Should it be taught in schools? The President revealed that he thought it would be a good idea for students to be exposed to the new ideas. His brief comments, made offhandedly in a wide-ranging conversation with reporters, were seized by the media and turned into a top news story. *Time*, which had been working on a story on Intelligent Design, suddenly rushed to complete its research and made the article its cover story the following week.[1] Authorities across the spectrum weighed in. Many bemoaned the scientific ignorance of the President, while others applauded his spirit of promoting free speech. Bush's comments set off a fresh explosion of public chattering and media debate on the pros and cons of teaching Intelligent Design in public school classrooms.

Lost in the shuffle was one crucial fact. The Discovery Institute, the central coordinating and funding agency for research on Intelligent Design (ID for short), had urged schools *not to require the insertion of ID into public school curriculum* because the theory is in its adolescence and any "teach-ID requirement" inevitably turns the question of design in nature into a political football. Rather, Discovery's scholars urged schools to revamp their teaching about the dominant theory, Darwinian evolution, so that negative evidence is no longer systematically excluded. In other words, *teach more about Darwinism than ever before*: teach the theory as is currently done, but also point out where it struggles with conflicting lines of evidence.

The day after reporters pried the brief comments out of the President and triggered the media frenzy, one of the leading opponents of ID, Paul Gross, appeared on the *O'Reilly Factor* television program. He said, "Intelligent design is a complex, highly proliferated body of action, literature, mostly PR, the purpose of which is to teach, or at least suggest, that there is a big body of scientific evidence showing that standard evolutionary biology is wrong, *that so-called Darwinism has collapsed*. That is all false."[2]

In mentioning the "big body of scientific evidence" that some perceive to have led to the collapse of Darwinism, Gross surfaces a key issue that generates a number of questions: Is Darwinism done for? Is it slowly spiraling downward into an unprecedented spectacle of global scientific collapse? Have its scientific and philosophical foundations truly cracked and crumbled beyond repair, as is argued vigorously by the scientists working within the Intelligent Design Movement? Is a new Intelligent Design paradigm emerging that retains Darwinian ideas only at the modest level of microevolution—variation of existing structures?

Or is it the other way around? Is Darwinism, subjected to powerful critiques from Intelligent Design theorists, emerging stronger than ever? In the wake of the rhetorical bombing and strafing that ID endured from scientists and the media over the past decade, is it ID, not Darwinism, that is collapsing under the weight of scientific criticism and "overwhelming evidence" of Darwinian evolution?

This double burst of questions captures the spirit of a great scientific clash that has broken into public view in recent years. This conflict is different from earlier versions of the endless debate over origins. *Now, the book of Genesis is not the issue.*

I am aware that Judge John E. Jones's decision in the *Kitzmiller v. Dover School Board* trial in December 2005 declared ID to be "not science" but rather a religious offspring of biblical creationism. This controversial decision, celebrated as a "2005 Christmas present" by Darwinists, has begun to boomerang on the Darwinist camp because of Judge Jones's egregious factual errors and his silence about the days of scientific testimony that quietly savaged the earlier testimony by Darwinian witnesses. Lehigh University biologist Michael Behe's published response alone pinpointed twenty serious errors—just in the science section of Judge Jones's opinion.[3]

I see the Dover decision as a fascinating footnote to a radically transformed debate about origins. Now there is laser focus on a specific set of scientific discoveries that are driving the new movement. Simply put, some researchers are arguing that as new layers of complexity are revealed in living systems, these hypercomplex, information-rich systems are straining faith in the Darwinian model beyond the breaking point. A typical two-paragraph summary of ID might sound like this:

> Scientific tests now show a shockingly severe limitation on the ability of random mutation to evolve new functional genes.[4] Also, the more we learn about the threadlike DNA molecule, which in human cells has 20,000 genes—digital files embedded on the cell's DNA hard drive—the more we realize that this DNA information is structurally identical to the ordinary coded information in human communication (books, digitized DVDs) and artifacts. To pin down what kind of cause "wrote the DNA files," we are able to apply a powerful reasoning approach that scientists now use called "inference to the best explanation." Since DNA (with RNA and proteins) have a mathematical structure called "specified complexity" (even one gene displays an astoundingly low probability, while its letters are highly specified), that enables us to ask a key question. In the real world, the world of scientific testing and experience, do we ever observe natural processes producing this kind of complexity? In fact, we have never recorded an instance where nature crafted this kind of complexity. *Yet, in the cause-effect structure seen in our world today, intelligent causes easily produce this kind of specified complexity. So the inference to design for DNA is based on our experience of the observed structures of the real world, not an imagined one.*[5]
>
> One finds equally compelling evidence for design in the bacterial flagellum, whose rotary motor drives certain bacteria through liquid like a submarine with an outboard motor. The flagellum, as biologists Michael Behe and Scott Minnich have shown, has a machinelike *irreducible complexity*, which is an empirical marker of design because it rules out step-by-step evolution through selection. Take one part away from the flagellum, and its rotary system won't work. Darwinian accounts of the evolution of the flagellum are (at best) sketchy "Just So Stories." Its forty parts, all of them precisely shaped proteins, are prima facie evidence of an intelligence behind life, and the flagellum is just the tip of the iceberg. The cell is chock full of such complex, multipart systems that continue to defy a step-by-step Darwinian explanation.

Of course, if a strong Darwinist (one familiar with the ID debate) read this paragraph, he or she surely would be loudly objecting at this point: "What about Kenneth Miller's critique of irreducible complexity? How can anyone buy ID's pathetic 'argument from incredulity'? How do Design theorists account for poorly designed systems like the human spine—or especially the human eye? How would a wise creator produce bumbling products like those?"

I know that if I were in the Darwinist's shoes, my mind would be popping with thoughts like these. My imagined cluster of responses reveals the *highly scientific adversarial nature* of the epic struggle between ID and Darwinism.

We have moved light-years beyond the stereotyped *Inherit the Wind* clash between dogmatic religion and enlightened science, which etched a fictional rendering of the Scopes trial onto our consciousness. Now, it's no longer William Jennings Bryan against Clarence Darrow—it's no longer religion versus science. Today it is ID biochemist Michael Behe of Lehigh University versus Kenneth Miller, Darwinian biologist at Brown University. Now it is ID theorist Scott Minnich, who teaches microbiology at the University of Idaho and publishes his research on the flagellum, engaged in intense discussion with Robert Pennock, a Darwinian philosophy professor who teaches at Michigan State University and has published critiques of ID. Whether anyone likes it or not, it is no longer science versus religion; *it is now science versus science.*

A Global Phenomenon?

Another sign that the ID controversy is not just a replay of the Scopes trial is the brute fact that this debate is spreading rapidly across the globe. Newspapers in Europe are now reporting the "dangerous" new ID concepts emanating from the U.S., and they are warning their people to be braced for this invasion. The penetration of Europe was symbolized by "Darwin and Design: A Challenge for Twenty-first Century Science," a conference held in Prague, Czech Republic, in October 2005. I was privileged to attend this gathering—the first major ID conference ever held in Europe. It drew seven hundred participants from eighteen different countries to hear pioneers such as Stephen Meyer, Jonathan Wells, and Charles Thaxton. Yet many commented on the participation of key speakers from Europe: John Lennox, a mathematician from Green College, Oxford University, who gave the final lecture; Dalibor Krupa, a leading physicist from the Slovak Republic; Cees Dekker, a world-renowned biophysicist who has pioneered biological nanotechnology at the University of Delft in the Netherlands; and David Berlinski, a philosopher of mathematics and science from Paris whose only religion is to "have a good time all the time." In my view, these four European participants gave presentations that totally obliterate the charge that ID is just "religion posing as science."

The Prague conference did not just have a strong scientific tang; it also had a distinct European flavor. Thus, what we are seeing today is not just a U.S. debate, mired in its own hypersensitized environment, which assumes that every questioner of macroevolution has religious motives. This new debate has leaped international walls; it is going global. It is also cross-disciplinary, and it is intensely empirical and mathematical, driven by the newest discoveries about the complexity and informational richness of nature. Several questions arise when confronting this global ID phenomenon: How strong are the arguments and evidences on each side? Who are the key players on each side, and what progress have they made in this fierce engagement with each other? What are the steps or stages through which the debate is developing, and where does it seem to be heading in the future?

This book is an attempt to answer these questions and more. The reader is invited to join me as I trace the current struggle between these two scientific perspectives—Darwinism (technically called neo-Darwinism since the 1940s)[6] and Intelligent Design theory. I shall be scrutinizing their intense clash in the 1990s and in the first decade of the new millennium. This saga—a complex and proliferating struggle of scientific persuasion—has now generated a high level of interest among scientists and within the general public. Beyond all of the basic factual questions mentioned above, we all want to probe deeper issues. What ultimate conclusions can we draw about our origins, based on the scientific evidence and on sound scientific reasoning? What is science, and what modes of scientific reasoning make sense?

In choosing the title *Darwin Strikes Back*, the focus is not so much on the early stages of the rise of ID as an idea and a movement nor on the opening stages of making the case for ID. That fascinating story is found in my earlier book, *Doubts about Darwin*, and in other books and articles.[7] This book is a deliberate sequel to my earlier work. It recounts how Darwin, incarnated in his modern heirs and defenders, has struck back furiously at the early inroads made by ID. It surveys the proliferating efforts by today's Darwinists to "crush the rebellion" (to echo the emperor's words from the *Star Wars* movies). It also highlights the energetic responses and countercritiques coming from ID theorists, as they use Darwinists' attacks to vindicate their own arguments.

A Personal Word

I probably should say a few words about my own bias as a historian of ID, working in the field of the "rhetoric of science."[8] I can relate to those who have ferociously attacked ID. My first encounter with anyone claiming "scientific problems with evolution" was an emotionally intense discussion at dinner in the fall of my freshman year at Princeton in 1968. To picture my mind-set that night, let me explain that as a teen I had declined in theistic belief from a vague deism during my youth to hard-core agnosticism by the time I was a senior in high school. I still attended church with my family, but it meant very little. My *God substitute* at that time, something

I could absolutely trust in, was *science*. I was an astronomy nut (I still am) and had written a term paper on the *big bang*, which wowed a high school science teacher. I was utterly convinced that—whether God existed or was a myth—one thing is undeniably true: we and all life-forms have evolved from a common ancestor. In high school biology, I was captivated by the concept of natural selection, which I took to be the most important law of nature ever discovered. In it, the inexorable creative power of nature is seen as it ceaselessly selects better life-forms. When I awoke to natural selection's power to create, it was an epiphany. I was just as Darwinian (and as committed to scientific naturalism) as Richard Dawkins, and I was not prepared for what I heard while eating supper with my Princeton friend John Donahue. He mentioned a study on origins he was attending, and I perked up my ears when I heard *evolution*. I asked about the study, thinking I might want to attend. John said that the teacher was presenting scientific evidence against evolution.

Donahue's words triggered shock and anger. "Evidence against evolution? There isn't any!" I blurted out. "Everyone knows that all the evidence supports evolution. Who is teaching this garbage?"

Shaken a bit, John told me that the study was presented by a Princeton alumnus from the class of 1913. I promptly pointed out that this gentleman would have entered Princeton when Woodrow Wilson was president of the university. This alumnus's problem was simple—he didn't know twentieth-century science! (As Richard Dawkins said: "If you meet anyone who doesn't believe in evolution, that person is ignorant, stupid, or insane.")[9]

I determined that night to meet and politely refute the ignorant alumnus. I met him at a public lecture on campus. We engaged in intense discussion that night for three hours and again at his apartment the next day. We entered into a quiet stalemate, and that led to other conversations with other Princeton students and alumni on the topic of God and origins. After six months of these discussions, including a series of weekly meetings with one young alumnus, I concluded two things: (1) I was not budged one inch from my belief in evolution, yet (2) my agnostic worldview was based more on hearsay and ignorance than careful research into the relevant evidence. By late May of 1969 I became persuaded about what C. S. Lewis called "mere Christianity," but my belief in evolution held firm—much as Lewis's acceptance of evolution was unshaken during most of his teaching career at Oxford and Cambridge.[10] In short, I was unmoved from my scientific beliefs by anything I had heard.

As a theist, I made it clear that I still found evidence of Darwinian evolution decisive. For some time, my position was one of a convinced Christian Darwinist. (By the way, evolutionists not only tolerate Christian

Darwinists—they are practically celebrated as trump cards in the debate with ID!)

It was only after many months (and years) of restudying the evidence, without my prior naturalistic bias, that I began to notice an anomaly here, an unanswered question there. The deeper I probed, the more I encountered implausibility upon implausibility in the story of macroevolution driven by nature alone. I was shocked to find out how weak the fossil evidence is supporting macroevolution of the phyla. By the time I received my degree from Princeton, I was convinced that microevolution (survival of the fittest) is solidly factual, but macroevolution (arrival of the fittest) was far less established on the foundation of fact.

During those years I was persuaded not by religious arguments but by scientific data. This same pattern of persuasion holds true for Michael Behe, Phillip Johnson, and virtually every leading light of the Intelligent Design Movement. *All were convinced by clear scientific arguments, based on empirical evidence.* When agnostic geneticist Michael Denton released his crucial 1985 book, *Evolution: A Theory in Crisis,*[11] I was further persuaded that evolution by natural selection was indeed a "paradigm in crisis," and that the unintelligent causes we find in nature can tweak existing structures but they cannot generate the complex motors or the vast genetic databases at the root of cellular life.

When ID began to emerge in the mid-1980s with the writings of Michael Denton and others, I felt I finally had found a scientific home—an intellectually satisfying approach to origins. Several things attracted me. One was the commitment to the highest standards of scientific quality. A second was ID's attempt to lower the heat in the tone of rhetoric—to avoid the bashing mode of discourse. A third was its central concept, which logically separated the *inference to design* from the separate task of *identifying the designer*. Here is how I explained it in a recent debate with Darwinist Michael Ruse: There is no "Made by Yahweh" engraved on the side of the bacterial rotary motor—the *flagellum*. In order to find out what or who its designer is, one must go outside of the narrow discipline of biology. Cross-disciplinary dialogue must begin with the fields of philosophy, sociology, history, anthropology, and theology. Design itself, however, is a direct scientific inference; it does not depend on a single religious premise for its conclusions. As I shall explain in the pages that follow, this conclusion seems compelling, unless one erects a rule excluding the design possibility as off-limits for consideration. If there is no such rule forbidding the consideration of design, it remains a live option. At that point, what matters is evidence and logic, not a preferred philosophy.

I certainly acknowledge that I have a personal bias, as does Richard Dawkins and everyone else who writes on this topic. I am a Christian theist, and that's all that matters in identifying my own bias. I am also convinced that naturalism (the belief that "nature is all there is") is much more problematic a bias than theism. Why? Simply stated, the question, "Did we and all life-forms arise through a long, gradual process driven by nature?" *is quickly settled at the level of one's worldview* if one simply accepts naturalism's prime catechism, "Matter gave rise to mind." If it is true that a *preexisting intelligence* is inherently mythical and not even possibly factual, then Darwinism (or something like it) wins automatically, no matter how weak the evidence. On the other hand, if various mind-first perspectives (including theism or deism) are *possible working frameworks of thought*, then the question "Did we evolve?" can no longer be simply settled at the level of one's worldview. At that point, one has to go further, wading into the evidence itself. And that's where we are headed: Where does the evidence lead us?

Una Mar de Gracias

I want to recall a lovely phrase I learned from my beloved high school Spanish teacher, Ruth Ferguson: "¡Una mar de gracias!" meaning, "A sea of thanks!" My indebtedness goes beyond a sea; I want to express an ocean of thanks to those who made this book possible. First to my wife, Normandy, who was steadily patient while I buried my nose in books or holed up with my laptop. Second, to the ID scientists who shared their thoughts and experiences, including Michael Behe, Walter Bradley, William Dembski, Cees Dekker, Robert Disilvestro, Robert Kaita, Dalibor Krupa, John Lennox, Jed Macosco, Stephen Meyer, Scott Minnich, Glen Needham, Paul Nelson, Ed Peltzer, Jay Richards, Charles Thaxton, Jonathan Wells, and Mark Whalon. Third, I'm also grateful for the interactions with scientists outside of ID, especially Richard Sternberg, an evolutionary scientist who suffered greatly for following normal procedures in allowing an article to be considered for publication (see sidebar on p. 27). I tried to interact with as many Darwinists as possible. I won't be naming them—lest they be charged with helping the ideological enemy! Yet I am grateful for their help. Fourth, I thank our wonderful friends Ron and Janet Vasquez, Loyd and Leslie Cunningham, Jerry and Ruth Swift, the board and friends of the C. S. Lewis Society, and others who helped quietly behind the scenes with their encouragement. I appreciate Bradley Jones's steady encouragement and helpful suggestions. Lastly, I thank my colleague Rich Akin who helped me with early drafts, and Chad Allen, my faithful editor. "Una mar de gracias para todos!"

Finally, this book is dedicated to my four children—Daniel, Stephen, Joy, and Karyn, along with their spouses and children. These are the rising generation who will grapple with Darwinism and ID and who will decide which of these theoretical models has proved itself the most fruitful explanatory paradigm for the future. May they come to love science for the wonderful, unfettered adventure that it is. May they think clearly and dare to question relentlessly, until they achieve answers that can withstand the closest scrutiny. Confident that they will do so, I can see science's greatest days ahead.

1

The Explosion of Design

"The Sky Is Falling!"

It was a surreal window in time. Beginning in August 2004, and stretching out over a year to the fall of 2005, the insidious threat spread across the globe. Month by month one could hear in the American media the staccato of increasingly shrill warnings. Editorial writers thundered across the land; journalists were scolded for inadequately reporting the danger on the horizon. Images of an impending catastrophe were conjured.

Then Oxford University Press joined the chorus, releasing two books that pinpointed the individuals who were linked to the new international threat. Cultural devastation, said the Oxford Press authors, was now lurking as a real possibility in the West. This was no small-scale matter—at stake was nothing less than our democratic values inherited from the Enlightenment. *Scientists and ordinary citizens needed to wake up and combat the menace; the health of our modern civilization was at risk.*[1]

This scenario sounds highly fictional—like a novel in which unspeakable terrorist plots against major cities are rumored and finally brought to light. One is reminded of the hype of a movie screenplay in which an approaching comet or asteroid is found to be on a collision course with planet Earth. Yet this *year of alarm* was not fiction. It was painfully real,[2] and when the seething controversy exploded in August 2005—triggered by an offhand comment at the White House—millions of Americans shook their heads, either in disbelief or in anger, as it was discussed in headline news and network newscasts.[3]

Blamed for the growing crisis was an unlikely group of troublemakers, most with Ph.D.s listed after their names. This scattered group in recent years had grown into a network of several hundred scientists and other scholars, many of whom were quietly toiling away in college classrooms and university science labs. Though cheered on by many in America and around the world, they suddenly found fingers of accusation pointed at them by leading spokespersons in academia and the media. In case you hadn't guessed it, the group bore a name: *the Intelligent Design Movement.*

These researchers were astonished to find themselves—and their unorthodox hypothesis of the design of certain features of the universe—thrown into the glare of public scrutiny. The topic was discussed on *Larry King Live* and in television specials, and it was analyzed in a *Time* cover story and a *USA Today* spread. In a more substantial vein, the theory was pummeled in a stream of hundreds of hostile articles and editorials and a dozen critical books. Criticism of their ideas was only to be expected, but several aspects of this flow of words shocked members of the Intelligent Design Movement. First was the high level of *contempt and hostility* directed at their point of view. Second, and equally astonishing, was the pattern of *crude distortion of their message and their motives*—with the worst often coming from fellow academicians. Advocates of ID could hardly believe their ears as they beheld their published critiques of Darwinism twisted beyond recognition, over and over, then dismissed with condescension as "nonscience." Worst of all, they found themselves *accused of spreading dangerous misinformation and endangering the health of science and even our very civilization*. An MSNBC writer, Ker Than, voicing the apprehension of Cornell historian of science William Provine, said that if ID successfully penetrates schools and universities, it will "become the death of science."[4] Those who worked under the banner of Intelligent Design found themselves thrust abruptly onto stage center of the cultural and scientific history of the new century. But they were treated not as scientific revolutionaries or respected dissenters but as public villains.

Two central questions arose quickly during the awakening of America to the Intelligent Design controversy: (1) *Who exactly were these controversial scholars?* and (2) *What led them to question the scientific creation story of life on Earth?* Many of the leaders in ID were tenured professors, and a number were considered pioneers or leading figures in their respective fields of research.[5] Some taught science or engineering at elite private universities such as Princeton, Yale, Oxford, Cambridge, and Dartmouth, while others labored in biology or chemistry labs at large state universities, including Michigan State

University, the University of Wisconsin, and the University of New Mexico. Scientists abroad had even added their weight to ID, including Dalibor Krupa, a physicist and member of the Slovak Academy of Science, and Lev Beloussov and Vladimir Voeikov, Russian biologists from Moscow State University. Beloussov, an embryologist, and Voeikov, a professor of bio-organic chemistry, are both members of the Russian Academy of Natural Sciences.

As to their motives, I focused on this question in *Doubts about Darwin* (2003), in which the early history of the Intelligent Design Movement was traced from the early murmurs of the 1960s to key developments at the dawn of the twenty-first century. One thing became clear from that review of the historical facts. Contrary to widespread allegations, *ID was not driven by a conservative Christian religious agenda*. In fact, the Ad Hoc Origins Committee (a forerunner to ID), far from being a gaggle of fundamentalists, was a very diverse group that was drawn together first by their skepticism of Darwinian doctrine but also by a general dissatisfaction with the approach of scientific creationism with its constructing of scientific arguments to support a literal reading of Genesis. Most of the members of the Ad Hoc group were not Genesis literalists, and some in fact were openly agnostic.

Denton and His Successors

What *was* crucial in the birth of Intelligent Design was a pair of conceptual bombshells—two key books that burst into public view in the mid-1980s, detailing the implausibility of evolutionary creation stories. These books began to build a shared structure of skepticism across the globe. They galvanized the forerunners of ID and shaped the emerging movement. The more explosive of the pair was *Evolution: A Theory in Crisis*, a 360-page manifesto by molecular biologist Michael Denton. When the book was published, in England first (1985) then the U.S. (1986), Denton was an agnostic geneticist, born and trained in England but working at a hospital lab in Australia. His book strives to amass sufficient data in every field of biology to crush the credibility of large-scale evolution. That is, while affirming microevolution, Denton contends that *macroevolution was certainly not a well understood process, and there was no evidence that it was driven along by mutations and natural selection*. While he emphatically rejects any return to a Genesis-based cosmology, he ends his book with a shocking assessment of the scant evidence for evolution: "Darwinian evolution is no more nor less than the cosmogenic myth of the twentieth century."[6] Whatever the true explanation may be, says Denton, *it certainly is not in our possession now. We need to go out and discover it.*

It was the reading of Denton's book that instantly bulldozed the mild Darwinian beliefs of Michael Behe (a Lehigh biologist) and Phillip Johnson (a senior law professor at UC Berkeley), not to mention its impact on many other academicians who would join the ID Movement. Their reading of *Evolution: A Theory in Crisis* was the turning point. It led directly to their launching of their own research programs into Darwinism. Johnson's scrutiny of evolution began in his sabbatical in England in 1987, when he read simultaneously Denton's shocking critique and Richard Dawkins's vigorous defense of Darwinism, *The Blind Watchmaker*. After four years of research and writing, and after submitting his work to scores of biologists and other scholars for critical review, his efforts eventually yielded *Darwin on Trial* in 1991. Johnson added an epilogue in a 1993 revised edition, to respond to the firestorm of criticism he endured from many quarters, including a lengthy attack published by Harvard evolutionist Stephen Jay Gould. He has since added five other books on Darwinism or its naturalistic foundations.[7]

In 1996 Michael Behe practically eclipsed Johnson with his bestselling *Darwin's Black Box*, an investigation of many complex molecular machines that, he argued, defy any plausible explanation as to how natural selection had assembled them, step-by-Darwinian-step. These many machine parts (tiny, precisely shaped proteins) were all needed to achieve the existing function; take one away, and the function shuts down. Thus, the story of their gradual production over time seemed to rest on a leap of faith rather than realistic scientific testing. By the late 1990s it became clear that Behe's work, and its notion of *irreducible complexity*, had become the scientific center post of the ID Movement.

If Denton's book was the main catalyst of early ID skepticism, it worked in tandem with another bombshell, *The Mystery of Life's Origin*.[8] *Mystery* contained a fairly technical critique of then-current theories of the chemical evolution of the first cell. Published in 1984, it tracked (and helped accelerate) the abandonment of the chance hypothesis of life's origins, which was envisioned as unfolding in an ancient chemical souplike mixture contained in an evaporating pond or an oceanic environment. *Mystery* was in sharp contrast with anything in the genre of scientific creationism—to the point that two well-known evolutionists, chemist Robert Shapiro and physicist Robert Jastrow, contributed blurbs for the cover. James Jekel, a professor at Yale University's medical school, said in the *Yale Journal of Biology and Medicine*: "The volume as a whole is devastating to the relaxed acceptance of current theories of abiogenesis [chemical evolution]." Within a few years of its publication, two of *Mystery*'s three authors—Walter Bradley and Charles Thaxton—along with the writer of their foreword, former chemical evolutionist Dean Kenyon, were all three working together, building a scientific alternative to the prevailing notion that some "undirected process in nature" had produced life.

To sum up, what glued the diverse group of Intelligent Design advocates together was not a common religious crusade, although most were probably Christian or Jewish theists. Instead, the essential core of ID has been a *shared profound skepticism of the received wisdom of biology—a skepticism that grew year by year as they interacted with evolutionists*. In a brief survey of ID leaders in 2000, I was surprised how many reported that after their encounter with evolution's evidentiary problems (as brought out in the two books or later in Johnson's critiques), they experienced a "scientific conversion" and felt the Darwinian explanations had collapsed; they were simply no longer tenable. Thus ID was born out of intense discussion of the empirical problems of current scientific theory. The focus of early discussions of the Ad Hoc Origins Committee rarely, if ever, turned to the cultural implications of a Darwinian worldview. Rather, conversations pivoted on the empirical data, and secondly, how to frame new rules of reasoning that would permit a *responsible and rigorous inference to design*.

To persuade scientists to consider the possibility of an intelligent cause was a major task—and yet after the rise to prominence of Phillip Johnson in 1991, it became a closely related project of ID. Design theorists confronted a key roadblock to this new "unfettered" science: *the prevailing philosophy of naturalism (or materialism)*, which assumed that only natural or material forces and entities can be considered as possible causes in the history of the origins of the universe and life. On this point, ID viewed science as badly contaminated with a distinctly *theological* point of view: philosophical naturalism, which guaranteed to investigators that matter preceded mind rather than mind preceding matter and the complex, specified arrangements of matter. If anybody doubted that Darwinian science truly operated on such a religious assumption (the assured noninvolvement of mind in creating or shaping matter), Harvard geneticist Richard Lewontin helped lay such doubts to rest with his 1997 essay reviewing Carl Sagan's book *Demon-Haunted World*. Lewontin writes passionately about the "struggle between science and the supernatural," and when he uses the word *science*, he clearly means "matter-before-mind science" or "materialistic-naturalistic science." Lewontin says, "We take the side of science in spite of the patent absurdity of some of its constructs, . . . in spite of the toleration of the scientific community for unsubstantiated 'just so' stories, because we have a prior commitment, a commitment to materialism." Lewontin even acknowledges that "we are forced by our a priori adherence to material causes to create an apparatus of investigation and a set of concepts that will produce material explanations, no matter how counter-intuitive." Are there any exceptions, any limits to this material mind-set? Not according to Lewontin. He closes this section, "*Moreover, that materialism is absolute, for we cannot allow a Divine Foot in the door.*"[9]

Johnson and his colleagues argue that Lewontin's statement reveals a philosophical dogma—what I call a "built-in catechism"—which functions as a new type of Genesis-faith that simply decrees, "Matter preceded mind." If Darwinian science was built on such a philosophical or even theological construct (as seemed clear), then this construct deserved the most thorough and skeptical questioning. Nevertheless, to ID's opponents, this skeptical cast of mind was in itself a cause for instant suspicion and alarm. The matters of *common ancestry* and the *natural selection mechanism* were believed to have been long settled—why would anyone question established fact? And how can any scientist jettison naturalism—if it goes, couldn't any phenomenon be viewed as potentially "an act of God"? Yet far more dangerous in the sight of these critics was the second stage that ID theorists entered after 1996. They had gone beyond their doubts about Darwin and naturalism. They now said they were developing and testing a new theory, one claimed to possess greater plausibility in accounting for life's complexity. This theory was supremely controversial, of course, entailing either one of the most important advances in the history of science or one of the worst betrayals that science has ever faced. It was a theory that was not just open to the consideration of intelligent causes—*it was a theory that laid down principles and procedures for the reliable detection of design in physical structures.* The work of mathematician William Dembski was devoted almost entirely to the construction of a new detection system, a logical-statistical procedure crafted to detect where an intelligence had been involved in any physical object, phenomenon, or event.

Critics of ID were increasingly vocal in a point-by-point critique of these new ideas, but they scoffed even louder at the claim of design theorists that Darwinism was enmeshed in a paradigm crisis, as Denton had explicitly hinted in the title of his book and in the name of his final chapter, "The Priority of the Paradigm." The late philosopher Thomas Kuhn used the phrase "paradigm crisis" to describe an early, troubled stage in a genuine scientific revolution, leading finally to a "paradigm shift." These ideas were set forth in *The Structure of Scientific Revolutions* (1962),[10] one of the most influential academic books of the twentieth century. Some design theorists appropriated Kuhn's ideas and said that they were laying the groundwork for a new competing paradigm in biology.

To the critics of ID, the openness to *intelligent causes* combined with the attack on naturalism were little more than a subterfuge. An intelligent cause loomed as a not-so-subtle substitute for God. Furthermore, the reigning paradigm, neo-Darwinism, was said to be brimming with health.[11] In fact, the existing paradigm was seen as stronger than ever, supported by new evidence from fossils, molecular biology, and other fields. In the teeth of these developments, ID's moves were seen as plainly deceptive, slyly importing religion into science.

Most of all, the idea that design theorists were at the cutting edge of a paradigm shift was both infuriating to Darwinists and literally unthinkable. This appalling rhetorical invasion into their territory had to be stopped. Urgent action—*rhetorical action of the strongest sort*—was called for. It was time to awaken the scientific establishment to the real threat to "science as we know it." This call to arms could be heard in a gradual crescendo after 1997, but the trumpets blared loudly in a rare 2005 letter from Bruce Alberts, president of the National Academy of Sciences, to all the NAS membership, warning them of the encroaching danger: "I write to you now because of a growing threat to the teaching of science through the inclusion of non-scientifically based 'alternatives' in science courses throughout the country."[12]

This program of counterpersuasion by ID's critics was aimed at several key groups. At the top of the list were high school biology teachers and biology professors, scientists in other fields, and leaders in the media and politics. Yet the response to ID also was packaged for the general educated public in the U.S. and in other countries, where the virus seemed to be spreading. The goal was to convince those who were unfamiliar with Intelligent Design that the movement was simply based on religion, not science, and was not to be considered a *serious scientific alternative*. Far more hot and dramatic was the imaginative rhetoric of ID critics. A series of nightmares was painted, in which ID threatened the educational and scientific future of modern societies. Rhetoricians have a name for these imaginative constructions, mixtures of fact and faith: *fantasy-themes*. Normally, fantasy-themes function as collages of images and concepts; at the center are interwoven plot elements, usually including heroes and villains. In my earlier work, I proposed *projection themes* as a more congenial and accurate term, since such projections aren't complete fantasies. (I now use both terms interchangeably.)

The dark projection themes of Darwinists (e.g., "Science will die!") are among the most amazing and fascinating responses to ID. But more commonly the rhetoric aimed at ID took the form of sound bites and talking points on television interviews and in short editorials and opinion pieces. The thrust of these jabs at design was a series of poison-tipped generalizations:

> There is no evidence for ID's arguments.
> There is overwhelming evidence for evolution.
> The newest findings are confirming Darwinian theory each month.

ID is not testable.
Alternately, ID has been tested, and it utterly fails.
ID is a sham; it survives only as a political or religious movement.
ID is pseudoscience; it never publishes findings in peer-reviewed journals.

Some of these sound bites began to lose their force, especially when articles supporting ID began to appear in peer-reviewed journals in 2004. Most prominent among a flurry of such peer-reviewed articles was Stephen Meyer's August 4, 2005, review essay, "The Origin of Biological Information and the Higher Taxonomic Categories." It was published in a journal loosely connected to the Smithsonian, *The Proceedings of the Biological Society of Washington* (vol. 117, no. 2). (For NPR's coverage of the Smithsonian's shabby treatment of the journal's editor, see the sidebar "The Sternberg Affair" on page 27.)

Nevertheless, ID critics went beyond editorials and jabs and addressed the issues of scientific evidence and arguments for design in a more head-on manner in extended articles or even book-length reviews. Such books were sporadic in the late 1990s, but after 2000 there was a steady flow, which built to a crescendo during 2004 when four books were published that blasted ID. The typical message of most of the book-length attacks was simple: "The arguments of Michael Behe, William Dembski, and other ID theorists have fallen flat. They have collapsed under the weight of criticisms and refutations of all their key points." Authors trumpeted findings in molecular biology that were seen as tests and refutations of irreducible complexity. Michael Behe sat permanently in their crosshairs, and scientists argued that his complex machines and systems can be, and have been, evolved. Case closed.

Behe and the other ID theorists wasted no time replying to the major critics—in print and in web-published replies. Occasionally, as in William Dembski's *The Design Revolution* (2004), a book-length reply to critics was unleashed. Design theorists argued that the attacks on their work, far from refuting their arguments, *actually strengthened and vindicated the case for Intelligent Design*. Behe published several articles and book chapters in which he replied, point-by-point, to his critics and sought to turn the attacks into powerful confirmation of his own theory.[13] Perhaps more than anyone, Jonathan Wells found the "rave reviews" of his work (negative raving) as constituting a massive vindication of his *Icons of Evolution*. The seesaw battle of words, arguments, and evidences slung back and forth with increasing vehemence after 1997 was the most spectacular rhetorical clash to emerge in the twenty-year conceptual war over Darwinism and design.

The Sternberg Affair

On November 10, 2005, National Public Radio's Barbara Hagerty covered on *All Things Considered* the lingering brouhaha over Richard Sternberg, the editor of "an obscure scientific journal loosely affiliated with the Smithsonian Institution, where he is also a research associate." She pointed out that "he published in the journal a peer-reviewed article by Stephen Meyer, a proponent of intelligent design," and then fleshed out the motives and misery of Sternberg:

"Why publish it?" Sternberg says. "Because evolutionary biologists are thinking about this. So I thought that by putting this on the table, there could be some reasoned discourse. That's what I thought, and I was dead wrong."

At first he heard rumblings of discontent but thought it would blow over. Sternberg says his colleagues and supervisors at the Smithsonian were furious. He says—and an independent report backs him up—that colleagues accused him of fraud, saying they did not believe the Meyer article was really peer-reviewed. It was.

Eventually, Sternberg filed a complaint with the U.S. Office of Special Counsel, which protects federal employees from reprisals. The office launched an investigation. Ultimately, it could not take action because Sternberg is not an employee of the Smithsonian. But Sternberg says before closing the case, the special counsel, James McVay, called him with an update. "As he related to me, 'The Smithsonian Institution's reaction to your publishing the Meyer article was far worse than you imagined.'"

McVay declined an interview. But in a letter to Sternberg, he wrote that officials at the Smithsonian worked with the National Center for Science Education . . . and outlined "a strategy to have you investigated and discredited." Retaliation came in many forms, the letter said. They took away his master key and access to research materials. They spread rumors that Sternberg was not really a scientist. He has two Ph.D.s in biology—from Binghamton University and Florida International University. In short, McVay found a hostile work environment based on religious and political discrimination.

After repeated calls and emails to the Smithsonian, a spokesman told NPR, "We have no public comment, and we won't have one in the future."

This fast and furious campaign of attack by the critics of Intelligent Design, with the vigorous counterattacks by ID's defenders, is the primary focus of this book. I want to tell the story of this intense period and, in the process, to separate and trace the battle over each of the main arguments. Some are scientific, some are philosophical, and others hover at a popular level, in culture and educational issues. The key to understanding the rhetorical action is to tease apart the main threads, grasp the central points, sift the crucial evidence, but not get lost in minutiae. That is my goal.

In this survey I want to track the clashes over ID *after 1997*, as the movement ballooned and spread across the university world and grew into a major intellectual and cultural issue within the general public. Because I already focused in *Doubts about Darwin* upon the responses received by Phillip Johnson's critique, the spotlight will shift in this book to Michael Behe and William Dembski, the leading theorists of "design detection," and to Jonathan Wells, the critic of "textbook proofs" of Darwinism (such as the peppered moth story and the comparison of embryos). Chapters 5, 6, and 10 will be devoted to their work, the counterattacks they faced, and their replies to the detractors. Chapters 8 and 9 will discuss one of ID's strongest criticisms of current textbook orthodoxy—the proliferating chemical evolution (origin of life) scenarios.

The fossil evidence—both for and against Darwinism—comprises a major battlefront in the war of rhetoric. The clash over fossils will be dealt with in chapter 7. Since the parallel arguments to design from the fine-tuning of the universe continued to play an important and increasing role in buttressing the case for design in biology, I will also devote chapter 11 to discussing the struggle over these arguments and evidences in physics and astronomy, along with the curious use of theological arguments to shore up Darwinism. Finally, in a concluding chapter, I will seek to understand this moment in time and to project some likely changes and pathways to be followed in the coming years.

To trace the story of how Darwinism struck back at ID and opened a new phase of the debate, it would be wise to devote three brief introductory chapters to a historical overview of this period before entering the argumentative clashes. I will first focus on the basic conflict between the two theories and ask: How did the relatively modest claims of ID (compared to creationism) generate such intense hostility? What was truly at stake, and what exactly was the perceived threat from design? Then, during two historical survey chapters on the 1990s and on the period after 2000, I will retell this exciting clash between design theorists and defenders of the fortress of Darwinism. We now turn to those introductory stories and questions.

The Real Issue

Nature's Symphony of Macroevolution

The controversy over Intelligent Design places before us a central puzzle. On the one hand, ID theorists have made seemingly *minimal claims* (compared to the much more extensive claims from creation science, which include a global flood and recent creation). At its core, ID simply says that "certain features of the universe and of living things are best explained by an intelligent cause, not an undirected process such as natural selection."[1] On the other hand, by 2005 the central concepts of ID and the professors who articulated them were increasingly seen—and *portrayed*—as a dire threat to science and even to the health of modern societies. How could these minimal arguments become the trigger for such unprecedented alarm in science?

To solve this puzzle, one must realize first that these minimal claims were viewed tactically as much more dangerous than traditional creationism in three ways:

1. They seemed more likely to penetrate science, since they were promoted by credentialed scientists, many of them "intellectuals at respected universities."[2]
2. The claims themselves (with no hint of a literal Genesis) were often linked with those scientific fields already busy with the detection of intelligence. Thus the claims posed a greater threat of penetration.

3. In spite of the minimalism and greater penetration ability of ID, its threat was calculated as not just scientific but also cultural. In other words, it was virtually as dangerous to scientifically informed cultures as the openly biblical version of creationism. This cultural danger seemed clear, since the goals of Intelligent Design—as expressed by its official think tank, the Discovery Institute—included the dethroning of the reigning philosophy of materialism.[3]

Yet, in communicating the overall danger of ID to the public, such tactical points could not always be presented openly. If they were expressed at all, they were often muted and phrased carefully and even indirectly. The primary danger—in almost all anti-ID rhetoric after 1996—was said to be the encroachment of "bad science" that was proceeding relatively unchecked. ID's attack on evolution, though painted as totally wrongheaded, was nevertheless spreading and gaining converts. This trend did not bode well; it was viewed as tantamount to the subverting of science, and this perception fueled a deepening panic.

It followed naturally that the negative portrayal of ID's scientific position took the form of a harsh polemic that gave no credit to ID for even pointing out unsolved problems in Darwinism. Typical (actually on the mild side) was the opening comment of Michael Ruse in his 1998 article in *Free Inquiry*:

> To the working scientist, and not just the biologist, it is simply ludicrous to think that there is any question about the natural origin of organisms from forms very different—ultimately, from inorganic materials. This is as much a fact of nature as that the earth goes around the sun or that water is made from oxygen and hydrogen. . . .
>
> Recently, the naysayers have gained more authority as their ranks have been swelled by people of distinction and position—not biologists working on the problems that concern evolutionists, but from other areas of science. . . . I shall examine proposals that these critics have made as an alternative to evolution through selection, in particular, the pretensions of the supposedly new hypothesis about "irreducible complexity," a phenomenon that demands the invocation of a Supreme Being of some sort. This is a very old argument indeed. Far from being a genuine alternative to evolutionism, it is neither needed nor plausible. On its own terms, it is riddled with problems.[4]

Leaving aside the crude distortion of Behe's argument (his alleged "invocation of a Supreme Being" from irreducible complexity)[5] and passing over the obvious incoherence in Ruse's second paragraph conflicting with his comment about "working scientists" in the first, this opening attack was quite typical of the polemics from Ruse's codefenders of Darwinism:

they painted ID as a total failure—the argumentation was flawed, and its scientific centerpiece, the arguments of Behe, were utterly without merit. The scientific counterarguments of ID's critics, supporting these charges, remained the strongest points hurled at this new threat.

The Heart of the Matter

According to critics, what were the crucial scientific failures of Intelligent Design? The charge of ID's "bad science" depended on how much space was available—whether a newspaper column, a lengthy article, or a book. Regardless of the length, there was almost always, stated or implied, a pithy accusation. Critics said that design theorists brazenly "rejected science" or "*had given up on science*." Such wording captured an imaginative universe of stories of scientific betrayal.

Typically, the construction of such betrayal scenarios employed several elements. First, design theorists who claimed to be unraveling the origins of complexity were accused by their opponents of turning their backs on science's core commitment to clear thinking and diligent searching. Phillip Johnson, for example, was accused regularly of not understanding "how science works." Michael Behe, on the other hand, was charged repeatedly with gross "laziness" in giving up on a search for solutions.[6] Second, ID theorists, according to critics, simply refused to follow the empirical evidence where it leads. Specifically, the new movement was turning its back on *overwhelming evidence* that has shown clearly that *Earth's creatures have changed over time*. (The word *overwhelming* was often wielded as a verbal sledgehammer to heighten the inexcusable nature of ID's error.) A third point was often added to stir speculation: Many evolutionary biologists have possessed religious faith and yet accepted evolution as God's chosen way of creating. So what's the problem with ID? Why can't these people get over their religious hang-ups with the scientific discovery of evolution? Are they not acting like religious extremists, practically surrendering their rationality, or at least blinding themselves to evidence, to protect sacred dogma? How can they deny nature's continual change that is staring them in the face? The bottom line was clear: Dangerous anti-intellectual attitudes had crept in and corrupted these scholars' normal reasoning ability. ID was *rejecting science*, and thus their pseudoscience must now be rejected!

Such underlying thoughts and images are the stuff of imaginative dramas, and they also flesh out the hidden subtext of much criticism of ID. Yet such fantasies and criticisms, woven and rewoven in many variations, have missed a large part of the story of the debate over biological design. ID's leaders, from the mid-1980s on, made it clear that they were not rejecting all of Darwin's theory. Michael Denton's arguments against Darwinism were built upon a careful distinction between small-scale evolution, or *microevolution* (fully plausible), and *macroevolution driven by natural selection* (overwhelmingly implausible, Denton argued). Every ID researcher and author since Denton followed in this same pattern of argument. All have acknowledged the basic scientific credibility of micro changes, or minor variations of existing structures, which will involve the origins of new varieties or even sister species. At this level, one can also track the true workings of natural selection to weed out the genetically unfit. Natural selection is real; it can be seen operating at this level.

So no one was denying *change over time*, the most vague and feeble definition of evolution, nor was there a dispute over *adaptations of species to their environments* or even *change in gene frequencies*, another popular definition of evolution. To behold biology is to see change over time. Species are observed ceaselessly adapting to the environment. For example, bacteria and insects *do develop resistance* to certain chemicals. Thus, in one sense there *is* overwhelming evidence of evolution—if you simply focus on the trivial phenomenon of microevolution. But before Darwin, the opposition to microevolution had already melted away.[7] Thus the key question remains: How does the evidence found for this limited kind of evolution (variation of existing form, through the tugging of natural forces) lead one automatically to extend the dotted line indefinitely to the production (innovation) of whole new kinds of organs and organisms? Is such an extrapolation legitimate? Most ID theorists argued that it is not.

One of the main strategies used against ID—with limited effectiveness—was to amass evidence for such minor changes and then to treat this as proof of nature's power to unfold the entire tree of life. An excellent example of this strategy came in November 2004, as more and more Americans were learning about ID. That month's cover of *National Geographic* posed a shocking question: "Was Darwin Wrong?" The article's first page repeated the question, but a headline on the next page shouted in gigantic letters, "No." The subtitle added: "The evidence for Evolution is overwhelming."

The article's author, Idaho freelancer David Quammen, uses a two-part strategy. First, he completely ignores Darwin's dissenters and their arguments. In effect, he pretends that ID doesn't exist. (This strategy is also seen in Stephen Jay Gould's final opus, *The Structure of Evolutionary Theory*.)[8] Quammen's second strategy to vindicate Darwin is crucial. He describes a wide variety of evidences, almost all of which can be placed inside a box labeled "microevolution." And yet this evidence, described as "overwhelming" and "persuasive," only overwhelms and persuades a person who makes no distinctions between microevolution—variation of existing structures—and the production of genuine innovation: new organs, new body plans, and new molecular machines in the cell. This latter sort of evolution—macroevolution—is the real battleground, but Quammen's piece is mostly silent on the aching questions that plague this realm of evolution. Notably, the ultimate macro-jump, which is never mentioned in Quammen's article, is the supreme mystery of all: the origin of life. How could the first living cells, requiring hundreds of genes, evolve from lifeless chemicals without intelligent guidance?[9]

Therefore, the crux of ID and of its real perceived danger in the eyes of critics was in the insistence upon separating microevolution from macroevolution and, secondly, in the relentless demand to see compelling evidence for nature-driven macroevolution. Although the official consensus in biology is that all major change was steadily sculpted by the laws of nature, ID argued that it was simply unwarranted to extend the very modest observed tweakings of organic parts into genuine innovation of whole new structures.

Nature at Work

Descending deeper into Darwinian thought, we see that the sole actor in macroevolution, according to mainstream biology, is *nature itself*, which is said to have the amazing power to mimic intelligence. Thus Darwinists viewed all the myriad forms of life today as the remaining links in a mostly extinct network of *unbroken natural chains*. To switch metaphors, the conductor of this symphony of life was the interplay of scientific law and chance; no real intelligence had ever made a detectable difference. At this point, we have descended all the way to the bedrock of Darwinism. The composer, who takes the place of creative intelligence in this drama, is natural selection. *Selection, we are told, is the engine of macroevolution; it has been shown to have fantastic creative power, ceaselessly reshaping life from one form to another and writing megabytes of DNA "computer code"—the tens of thousands of genetic files on the cell's hard drive.* From chemical broth, to pulsating bacteria, to darting fish, to slithering reptile, to yawning monkey, to theorizing man—all of this sweeping drama in the symphony of life was spun out by a master composer who doubles as the director with baton in hand. The composer's name (using Darwin's alias for selection) is "Survival of the Fittest."

This mechanistic theory of macrocreation through natural selection—Darwinism's "genesis story"—was vigorously and relentlessly attacked from all angles by design theorists; it was the Achilles' heel of biological evolution. Some theorists, Michael Behe most notably, do not even discard Darwin's tree of life—the common ancestry of all living things.[10] What Behe denies in *Darwin's Black Box* is that life's molecular motors and other systems of astonishing complexity were assembled by natural selection or any other unintelligent cause embedded in nature. What Behe affirms is that there is no good scientific reason, at this level of biology, to deny the inference that these machines were in fact designed by an intelligent agent. Science should be able to handle this shock, in spite of any metaphysical implications.[11]

The *ultimate perceived danger* emanating from ID theorists at this juncture (though this was not always stated openly) was the proposed reintroduction of intelligent causes into the explanatory toolbox of science. This idea constituted a profoundly radical departure from scientific convention. ID was publicly repudiating a doctrine that had dominated biology for a century and a half—since Darwin published *The Origin of the Species* in 1859. This doctrine, claiming that *only material forces* were responsible for the amazing diversity and complexity of life, was simply taken *as fact* since the late 1800s. The banishment of such possible causes as *intelligence* or even the philosopher's idea of *teleology* (things existing for a purpose) was viewed as a healthy purging for the sake of science's purity. "Good riddance!" was the nearly universal (though whispered) response of Darwinists.

This crucial Darwinian rule of reasoning—the mandatory exclusion of real design or teleology as the starting point for scientific investigation—had been given a name: *methodological naturalism*. This rule and its supporting philosophy became a battleground when Phillip Johnson wrote *Darwin on Trial*.[12] The ID heretics had analyzed and rejected this rule, asking what basis there was for restricting the possible causes that could be considered for any physical phenomenon. By rejecting the naturalistic rule, ID had crossed a line of acceptability within the dominant paradigm. They were paying dearly for this move from the 1990s onward—through a growing and vehement anti-ID campaign, which is the main focus of this book.

Resonating Arguments Outside of ID?

In spite of the publicly united front of all evolutionists in facing and opposing ID, the deep questioning of *macroevolution driven by mutations and selection* did in fact resonate with certain leading researchers—especially those in the field that is exploring how new body structures arose. To these evolutionary researchers, there were some widening cracks in the aging wall of Darwinian theory that were becoming unavoidably visible in the new millennium. One of the most important public signs that such problems were real and were being worked on with great intensity was a crucial gathering of a group of researchers affiliated with a hot field called "evolutionary developmental biology" ("evo/devo" for short). The meeting, held in 1999 near Vienna, Austria, sought a more plausible and comprehensive account of the "origination of organismal form." This carefully chosen phrase, which means essentially what this chapter has described as macroevolution, became the title of an extremely important book, *Origination of Organismal Form* (hereafter referred to as *OOF*), that grew out of the conference. Released in 2003 by MIT Press, the book was coedited by Austrian zoologist Gerd Muller and American cell biologist Stuart Newman, both leaders in theoretical biology.

Their point of view is clear from their introductory chapter, and it also emerges from many of the other contributed chapters from more than twenty different researchers. This radical point of view can be summarized in a sentence: the current, gene-centered theory can only reliably explain the diversification of body structures or forms once those forms have arisen; it cannot explain the rise (or origination) of those forms in the first place.

Of course there exists a crucial difference between the approach of ID and that of the researchers gathered near Vienna. The contributors to *OOF* assume that the answer to the question, "What drove macroevolution of new body forms?" will be a new, and probably more complex, *natural* theory. Presumably, the new sought-for theory will be an explanatory system that involves scientific law or laws, working over time, through the interplay of chance and necessity. To most authors in *OOF*, design theory presumably would be faulted with prematurely giving up on the problem of macroevolution. For many of these authors, ID was probably also seen as simply propounding a nonscientific answer, since it was going outside the realm of law and chance and pointing to intelligent causes.[13]

Yet in spite of this crucial difference, there is some significant common ground, and one can view the team of scholars that produced *OOF* as working parallel with ID, tackling the same huge problems, but using different approaches and guiding assumptions. One can almost imagine Michael Denton smiling and nodding in agreement with this vigorous new project, since it was, to some extent at least, vindicating one of the main points that he and ID had been making for two decades: the current neo-Darwinian theory of the origin of new body structures, supposedly arising from the random shuffling of bits of DNA over eons, is no longer a compelling or even plausible theory. To validate this radical parallel between *OOF* and ID, let me quote from the opening chapter:

> These developments [the dominance of evolutionary genetics and the overwhelming focus on the gene] have edged the field farther and farther away from the second initial theme: the origin of organismal form and structure.

> The question of why and how certain forms appear in organismal evolution addresses not what is being maintained (and quantitatively varied) but rather what is being generated in a qualitative sense. This causal question concerning the specific generative mechanisms that underlie the origin and innovation of phenotypic [i.e., bodily] characters is probably best embodied in the term origination. . . .
>
> That this causal question has largely disappeared from evolutionary biology is partly hidden by the semantics of modern genetics, which purports to provide answers to the question of causation, but these answers turn out to be largely restricted to the proximate causes of local form generation in individual development. The molecular mechanisms that bring about biological form in modern-day embryos, however, should not be confused with the causes that led to the appearance of these forms in the first place.[14]

The authors continue discussing the puzzle of macroevolution, pointing out that while natural selection is a force that plays some role in the evolution of new structures and morphology, the appearance of specific new elements of body construction "must not be taken as being caused by natural selection; selection can only work on what already exists." The parallel with ID's critique is obvious. The all-powerful role of selection in building genuinely new structures in nature is being openly questioned.[15]

In one memorable and revealing paragraph, the editors describe the macroevolution of new body structures as a "relatively neglected aspect" of evolutionary studies, and the failure "to incorporate this aspect represents one of the major gaps in the canonical theory of evolution." Indeed, a few pages later, they explain that the neo-Darwinian paradigm "completely avoids the origination of phenotypic traits and of organismal form. *In other words, neo-Darwinism has no theory of the generative*. As a consequence, current evolutionary theory can predict what will be maintained, but not what will appear."[16]

So What?

Thus ID's radical questioning of textbook macroevolutionary explanations, so bitterly opposed by evolutionists wherever such doubts were voiced, seems to be in fact an eminently valid line of questioning, if one listens to the scholars who wrote their exploratory thoughts in *OOF*. These theoretical biologists see the same problem that ID has focused on. Their tentative answers are profoundly different from design theory, but one thing is clear: Because of *OOF*'s candid assessment of a glaring mystery, the explanatory gaps in Darwinism are undeniable. ID is in fact pointing to a massive, unsolved problem in the existing paradigm.

Having clarified the core issues of the controversy, let us now track the contours of this ever-expanding conflict, beginning in the mid-1990s when Darwinists were still feeling aftershocks from Phillip Johnson's first two books, *Darwin on Trial* and *Reason in the Balance*.

Design after 1996

Advancing through Heavy Fire

In *Doubts about Darwin* I told the story of an impromptu, furious debate that erupted between Stephen Jay Gould and Phillip Johnson in front of ten other scholars who had gathered for a weekend meeting in Boston to discuss the "problem of evolution and creation in the public schools."[1] In the context of that Jedi fight with verbal light sabers crackling, I wove in the supportive role of one onlooker—a sympathetic evolutionist who played a key role as a friendly critic and fact-checker for Phillip Johnson. I'm referring to the leading paleontologist David Raup of the University of Chicago, who vouched for the accuracy of Johnson's scientific writing in front of the Boston gathering. In the fall of 2000, Raup told me that Johnson's work "is very good scholarship and, of course, this has been widely denied. He cannot be faulted; he did his homework and he understands 99 percent of evolutionary biology."[2]

The three key individuals in this story—Johnson, Gould, and Raup—symbolize three key types of scientific actors in the drama we're entering: the Besieged ID-Heretic, the Zealous Inquisitor, and the Courageous Inquirer. By necessity, this chapter will emphasize the first two types of actors, since they are the loudest and most visible protagonists on both sides of the battle lines. At the same time, I never want to overlook or minimize the important role of the Courageous Inquirers, whom I describe as open-minded evolutionists who (1) are painfully aware of the problems, anomalies, and even foundation cracks in the current Darwinian paradigm, and (2) are interested in bringing such heretical ideas as Intelligent Design into the discussion to see if some progress may come of it.

I must keep details confidential, but I have been permitted to sketch an amazing lunch meeting that took place recently at the faculty club at one of the world's most prestigious universities, where a renowned physicist who had become profoundly skeptical of Darwinism met for lunch with a leading evolutionary biologist with whom he had become acquainted. A primary goal of the physicist was to find common ground and come to a mutual understanding. In quiet tones, over a delicious meal, both of these scientists quickly found one scientific area they agreed on: the evidence is now compelling that an intelligent agent was behind the origin of the first cell. In other words, they agreed that chemical evolution had revealed itself as a dead end and thus a *designing intelligence* was now the most plausible explanation for the first life. Next, the physicist reported on his reading of Gould's 1400-page final opus, *The Structure of Evolutionary Theory*. The biologist listened intently as the physicist shared his observations and detailed why his doubts were only reinforced by reading Gould. Lunch ended on a positive note with no reported harsh words or indigestion. In this case, the ID-Heretic (the physicist) was not branded as such by his dialogue partner, the Courageous Inquirer (the biologist). Because of the commitment on the part of both professors to seek scientifically plausible explanations in an open and respectful dialogue, progress was made in mutual understanding.

I hope that such calm, rational encounters will be the real wave of the future. Sadly, the most prominent voices by far from mainstream biologists have not been those of Courageous Inquirers but of Zealous Inquisitors. The first phase of the ID-inquisition began as Phillip Johnson's two earliest books, *Darwin on Trial* and *Reason in the Balance*, began to invade the consciousness of university campuses and the American public in the period of 1991–1995. Johnson's writings and his frequent lectures on college campuses were probably the greatest cause of worry within the biological establishment during this period, but many opponents were working hard in the mid-1990s to deflect his arguments.

At the same time, the defenders of Darwin could not so easily ignore biochemist Michael Behe and his book, *Darwin's Black Box*, especially when Behe's pivotal importance and his new ideas began to permeate the U.S. and the world in 1996. Whereas Johnson's work rarely attracted major media attention, Behe's did—over and over, and at the highest levels. The focus on Behe was stronger than ever nearly a decade later when seven paragraphs were devoted to his work and his point of view in the *Time* cover story in August

2005. In retrospect, the rise of Michael Behe might be viewed as the first major advance of ID after the initial flurry of excitement over the critiques from Michael Denton and Phillip Johnson. I mark this new period as starting in the late spring of 1996 when Behe's publisher, the Free Press, prepared to churn out its first run of 10,000 copies of *Darwin's Black Box*.

The Big Picture

To frame the action in this period, let me pause to sketch the larger picture. The historical pattern that followed the release of Behe's book in late July of 1996 might be compared to an army's coordinated sweep up a steep mountainous terrain to capture a crucial fortress. Here the fortress represents the assumed plausibility of nature-driven macroevolution and the accompanying monopoly of Darwinism as the only accepted point of view in the academy. With the advent of Behe (and later, through the publications of Jonathan Wells, Stephen Meyer, and William Dembski), ID had suddenly begun to employ a new and more powerful set of weapons that enabled them to move more rapidly toward the fortress. With each advance, inching closer to their target, the defenders inside the fortress were repeatedly mobilized. This mobilization task was mainly under the leadership of Eugenie Scott, who served as director of the National Center for Science Education (NCSE) in Berkeley, California. She monitored all kinds of creationist movement into the educational arena so that it could be confronted and halted.

The defenders unleashed wave upon wave of heavy shelling, using their own bombs and rockets with the high explosives of scientific rhetoric. They maintained their hope that they could halt the advance and even repel the invasion entirely. Thus in this metaphorical clash, each major advance came when ID's case for design came to public visibility through media exposure or when hostile evidence or other problems with Darwinism came to light. Each new ID book contributed to this advance, as did the occasional symposia or conferences on university campuses. As heavy fire descended upon ID, progress was checked somewhat, and this occurred when pro-Darwinian scientific discoveries were announced or when opponents unleashed their own counterattacks through articles, books, and public presentations. For the rest of this chapter, I will track the major stages of ID's advance through heavy fire, focusing on the key moves on each side.

We first turn back to Michael Behe, whose string of publicity breakthroughs began with a bang on August 4, 1996, when the *New York Times* published a surprisingly positive review by James Shreeve of Behe's book. Shreeve praised Behe as a writer, took note of his interesting challenge to

Darwinian orthodoxy, but in the end came down on the Darwinist side with a "don't give up on Darwin yet" response. Shreeve expressed hope that scientists in our grandchildren's generation might unravel the natural pathways by which complex cellular machines were evolved. This line of argument became one of the most common replies to Behe—hinting (or stating) that his book sprang from a lazy attitude, giving up on the problem prematurely.[3]

Less than three months later, the *New York Times* invited Behe to summarize his argument in an opinion piece, "Darwin under the Microscope" (October 29, 1996). His column was published partly because the editor of editorials had enjoyed Behe's book and also because the Pope's apparent endorsement of evolution was splashed across the front pages of newspapers around the world, including the *Times*, a few days earlier on October 26. (This spin on the story was later found to be a distortion of his comments.)[4] The editor felt that Behe, a Roman Catholic, could bring a new and interesting perspective on evolution. Twice more in the coming years,[5] Behe wrote opinion columns for the *New York Times* and was featured positively in news reports, even though the *Times* made it clear that from an editorial point of view, they rejected Intelligent Design as a legitimate challenge to Darwinism.[6] Nevertheless, the *Times* made a momentous decision each time they allowed Behe to sketch his ideas about the cell's irreducible complexity and to explain why this evidence for design was unmistakable.

Behe's work was quickly recognized as raising the threat level against Darwinism by several notches, since he was writing in his own field (molecular biology) and was crafting what seemed to be a genuinely scientific argument for detecting design. His motives were somewhat less vulnerable than Johnson's. In Behe's case there had been no midlife religious conversion, only a scientific one. In fact, he capitalized on this point, incessantly reminding the public that it was the empirical evidence, not his lifelong Catholic religious beliefs, that captured his attention and propelled him as a Lehigh biology professor to doubt the Darwinian creation story. He explained that it was the multiple lines of anomalous data, as laid out by Denton and later Johnson, that drove him to research evolutionary biology and the evidence in the cell's nanotechnology that pointed to design.[7]

Darwin's Black Box rapidly took control of the conversation over the plausibility of macroevolution, focusing on the molecular-level realities where natural selection seemed, on the face of it at least, utterly incapable of such foresight and planning as are needed to build and assemble the many components of tiny intracellular systems. When Behe injected *irreducible complexity* and other provocative phrases and ideas into the academic discussion, it brought many of the world's leading defenders of Darwinian theory on the run to defend the Darwinian fortress against Behe's advancing platoon. Extremely unhappy over the level of attention the media had given Behe, these defenders launched several websites devoted to criticisms of his ideas. Some attacked the quality of his research, and Behe responded point-by-point to the early attacks. Dozens of critical reviews appeared. Before long, direct refutations of Behe's ideas were released, initially in the form of journal articles and magazine essays, arguing that Behe's meager attempts had "collapsed" (this seemed to be the word of choice in attacks on Behe).

Firing the Big Guns

The year 1999 can be marked as a turning point—the year that a major Darwinist counteroffensive began. Leading the way were two books that focused heavily, if not exclusively, on the threat of Intelligent Design. One came from biology professor Kenneth Miller at Brown University, a charismatic figure who became Behe's relentless opponent in the years that followed. Miller's book, *Finding Darwin's God*,[8] became a pillar of defense against ID. Its title connects with the fact that Miller, the bearded coauthor of a biology text used in U.S. high schools, was also a fellow Roman Catholic with Michael Behe. Thus Miller employed a hybrid approach, claiming to confront two abuses. First, all creationists (including Behe and ID generally) were standing on weak arguments and were thus illegitimately using science to point to God.[9] Second, in a key chapter that echoed what ID theorists had said all along, Miller scolded his colleagues who were outspoken dogmatic atheists, like Richard Dawkins, for their forceful assertion of atheistic implications coming from evolution.

Kenneth Miller found himself debating Behe on several occasions, including a four-person debate at the American Museum of Natural History in New York City. This remarkable two-hour event, moderated by Eugenie Scott, paired Behe with Miller (and Dembski with Robert Pennock, whom we will shortly meet). Design supporters viewed the results of this debate with great enthusiasm. It was seen as proof of their progress that Miller's arguments had not progressed since the earlier encounters and that Behe stood his ground well.

Clearly, Miller was the most polished and assertive rhetorician in the Darwinian fold of scientists who were prepared to speak out against ID. He spoke with bravado, displaying a brisk and impervious confidence in his own position. He never missed a chance to summarize a line of evidence, then swing his verbal sledgehammer, announcing (for example) the "collapse" of the ID position. A typical flash of Miller bravado is seen in "The Flagellum Unspun," from *Debating Design*: "The great irony of the flagellum's increasing acceptance as an icon of the anti-evolutionist movement is that fact that research has *demolished* its status as an example of irreducible complexity almost at the very moment it was first proclaimed."[10] In chapter 5, Miller's arguments will be listed and confronted to see if Behe may have managed to turn the tables on Miller.

Of the two anti-ID bombshells of 1999, the more massive one (almost four hundred pages of small print) was Robert Pennock's *The Tower of Babel*. He invented a clever new label: "Intelligent Design Creationism" (or "IDC"—note the tactical importance of Pennock's addition of the *C* word), and he located the IDC proponents within the larger pool of all creationists. One insightful comment focused on how ID has created a new "second front" in the struggle to penetrate universities. (Pennock says the first front is the indoctrination of high school students, by conservative churches, before they enter the university.) This second front is

> that many of the new creationists, unlike their predecessors who operated out of private ministries, have acquired positions in colleges and universities and are leading an attack from within. They work in tandem with campus Christian student groups to hold creationist conferences, and they arrange for departments to sponsor anti-evolution speakers. Their academic credentials and affiliations give them entrance to broader public forums, so, for example, one can now find a few of their books published by academic presses instead of only by small Christian presses. They rightly see this move into higher education as an indication that their movement has achieved a new momentum. Creationism is poised to break into the mainstream.[11]

Pennock spent a considerable portion of his book in a point-by-point attack on Phillip Johnson.[12] He also criticized Behe and Dembski. It is noteworthy that this book was the first major attempt from the side of Darwinian orthodoxy to identify all major players and arguments in Intelligent Design and to try to controvert each major ID point.

Pennock, a tall, tousle-haired philosophy professor at the University of Texas when he began his research, taught briefly at the College of New Jersey and later settled at Michigan State University. He had been tracking the Intelligent Design Movement virtually since its public inception. I have enjoyed several conversations with Pennock—most recently when he attended a debate on Intelligent Design that was arranged between computer science professor Donald Weinshank and me at Michigan State University. I found Pennock to be affable in dialogue, but he has proved relentless as an opponent of the Design Movement. Pennock's importance as an ID critic cannot be minimized; he played a leading role in organizing responses to ID and gave key testimony in the Dover trial in the fall of 1995.

Pennock was also notorious for editing a massive and controversial paperback, released by MIT Press in 2001 with a whimsical, bright yellow cover that sported a pair of rabbits. This volume, *Intelligent Design Creationism and Its Critics*, fully twice as massive as his own earlier tome (800 pages), is an oddity in the history of attacks on ID. First, though it seems at first glance to present all sides adequately, it contains a grossly imbalanced treatment of ID, allotting less than one third of the space to ID theorists. Consider another lopsided feature: twice in the book an ID theorist expresses his ideas or his critique of Darwinism, then not one but *three Darwinian responders reply*. In football, there is a name for this: "Piling on." Pennock, consistent with this weird imbalance, structures each of nine subsections of the book, except one, in such a way that a critic of ID is given the final word. Another way Pennock's "big yellow book" became notorious is its use of some published ID articles without asking permission of the author or even notifying the authors of the plan to use previously published works.

Pennock established in both of his books a basic strategy for anti-ID rhetoric. Intelligent Design, he argued, is merely a gussied up, slightly sophisticated version of *biblical creationism*.[13] ID was described as differing in only minor points with the older ideas of "scientific creationism"—that is, in using scientific data as support for the biblical accounts in Genesis. This attempt to link the two movements is seen throughout *Tower of Babel*. The not-so-subtle subtitle is *The Evidence against the New Creationism*. Without exception, defenders of Darwinism sought rhetorical advantage by this *lumping in* tactic; they branded ID with the mark of its hated and feared cousin—scientific creationism. Of course design theorists rejected this lumping as utterly baseless. It was tagged as "empty and manipulative rhetoric," since ID theory does not depend upon a single biblical or religious premise.[14]

Before passing on to the next phase of the battle, I will mention one more anti-ID book of somewhat lesser importance, released in 2000, just after Kenneth Miller's and Pennock's works: *The Triumph of Evolution and the Failure of Creationism*, by Niles Eldredge, Stephen Jay Gould's partner in developing the punctuated equilibrium model of evolution. This chatty book is to some extent a recycling of material used in his earlier work, *The Monkey Business* (1982). But Eldredge in his new book added several chapters on the newer, more dangerous creationists of Intelligent Design. This book is hobbled with major errors of fact and distortions of the ID position. For example, he says that creationists hate and oppose evolution, motivated by "the belief that evolution is inherently evil." (There is abundant evidence, beyond Behe's story, that ID is primarily motivated by intellectual contempt for the feeble arguments for macroevolution, not by fear or alarm over its "evil nature." Even relatively uneducated creationists are far more motivated by sheer *intellectual skepticism* than Darwinists care to admit.) Second, Eldredge stumbles badly in asserting that the majority of creationists "are motivated primarily to see that evolution is not taught in the public schools of the United States."[15] This is so flagrantly false, so clearly against the data, one wonders why an editor (or Kenneth Miller, who contributed a glowing endorsement) did not catch it. No major creationist organization has advocated such a position in recent memory, and even polls falsify this bizarre claim.

Meanwhile, British bacteriologist Alan Linton reviewed Eldredge's book respectfully but with strong reservations in the *Times Higher Education Supplement*. Under the catchy title, "Scant Search for the Maker," Linton closed with a ringing vindication of Behe's arguments: "The biochemical complexity of cascades of enzymes required to perform a single function in the cell is mind-boggling, and for a structure or function to be selected it must be functionally complete. [Synthesis of amino acids from simpler chemicals is cited in the book], but this synthesis is nothing compared with the complexity of a single protein enzyme, let alone a series of highly specialized enzymes functioning in a cascade sequence. Such irreducibly complex systems are of no selective value unless they are complete." Coupled with this interaction, which in effect repudiates all that Eldredge has written about Behe, Linton goes much further and dissents with the claim of macroevolution, based on data in his own field of bacteriology:

> Despite the conciliatory comments in the final chapter, the book's title is essentially emotive and provocative. Since most theories, if proven to be false, are rejected by scientists, Eldredge claims that, after 150 years, science has failed to disprove the theory of evolution and, therefore, "evolution has triumphed." In other words, the theory of evolution rests on the failure of science to show that it is false. Nevertheless, he believes the theory can be scientifically tested.
>
> But where is the experimental evidence? None exists in the literature claiming that one species has been shown to evolve into another. Bacteria, the simplest form of independent life, are ideal for this kind of study, with generation times of 20 to 30 minutes, and populations achieved after 18 hours. But throughout 150 years of the science of bacteriology, there is no evidence that one species of bacteria has changed into another in spite of the fact that populations have been exposed to potent chemical and physical mutagens and that, uniquely, bacteria possess extrachromosomal, transmissible plasmids. Since there is no evidence for species changes between the simplest forms of unicellular life, it is not surprising that there is no evidence

> for evolution from prokaryotic to eukaryotic cells, let alone throughout the whole array of higher multicellular organisms.[16]

This shocking radicalism of response from Linton will be echoed later in the book when we encounter the American biologist Ralph Seelke and hear what he has to say about the lessons learned from bacteria.

At the Millennium: A New Proliferation of Design

As the new millennium opened, there were several reasons for hope within the ID Movement, even though storm clouds hovered on the horizon. The powerful new impetus to ID from Michael Behe had been supplemented by two other breakthroughs: First, the outlines of foundations of a new paradigm seemed to be emerging in three important new works from William Dembski, released in 1998–1999.[17] Central to all three books, and to the sense of encouragement in his work, was Dembski's creation of the "Explanatory Filter," a logical and statistical procedure by which one could determine if an object, event, or system had in fact been designed by an intelligence. The fact that the filter was *conservative* (intelligence was inferred only after lawlike processes and chance had been ruled out) and that *it yielded a bare inference—"design by an intelligence"*—gave new promise to the ID notion that design can be positively detected apart from any necessary connection with theology. Dembski's work was, of course, met immediately with various criticisms by more and more Darwinists, some of whom we will meet in chapter 10. These came from both credentialed scholars as well as a much larger army of Internet enthusiasts of Darwinism—the same general group who had leaped to criticize Behe's work. In spite of the difficulties Dembski faced (such as his struggles at the Polanyi Center at Baylor University),[18] there was an overall sense of ID Movement encouragement, boosted further by Dembski's releasing another book on his mathematical arguments for design, *No Free Lunch*.[19]

The second great encouragement at the dawn of the millennium was the book *Icons of Evolution* by Jonathan Wells, a recently minted Ph.D. in molecular biology at UC Berkeley. His book was a pointed embarrassment to Darwinist "proofs" of macroevolution and the chemical evolution prelude. Each of ten selected icons (pictures or diagrams prominent in high school biology texts that illustrated the truth of evolution) was found to be laced with misinformation, misleading statements, omissions, and even borderline fraud. This book, and the storm of criticism that it sparked, is the focus of chapter 6.

The third and last significant uplift ID received at this millennial turning point was the sense of momentum gained through the inception of conferences across the country that featured ID on university campuses and even the launching of a network of college clubs—IDEA Clubs (or Intelligent Design and Evolution Awareness Clubs). The first IDEA was established at the University of California at San Diego by Casey Luskin. By 2005 chapters had spread to dozens of other campuses, and this was seen as such an alarming development that even *Nature*, the world's most prestigious science journal, featured the IDEA phenomenon in a cover article in the issue of April 28, 2005. The article, by Geoff Brumfiel, was entitled "Who Has Designs on Your Students' Minds?"

The proliferation of ID-oriented clubs was a long-term trend, but there were also several larger conferences whose impact was hard to calculate. In *Doubts about Darwin*, I discussed "The Nature of Nature," the Baylor Conference in the spring of 2000, organized by William Dembski and featuring a number of world-renowned scientists. Yet in the same general time period, two other conferences were held that made a historical mark. One entitled "Intelligent Design and Its Critics" was organized at Concordia College in Wisconsin by a British-born philosopher, Angus Menuge. This was significant in the long run because it yielded the extremely important book from Cambridge University Press, *Debating Design*, edited by Michael Ruse and William Dembski. Menuge's own role as organizer was honored by his being given an opening chapter, a brilliant short history of the ID Movement.

The other conference, held at Yale University in November of 2000, was seen by many as another turning point for not only public exposure but scholarly credibility of ID as a young but legitimate program in science. This conference, at which all the major ID leaders spoke, was left buzzing by a shocking new thesis presented by Guillermo Gonzalez, a young NASA researcher who taught at the University of Washington. In a new twist on the *fine-tuning* concept, which had been debated by physicists since the 1980s, he explained that planet Earth seems not only designed and fine-tuned for life, *it also appears to be fine-tuned for the benefit of scientific measurement and discovery*. This novel idea set conferees' minds reeling with the implications and possibilities. It was another indicator of the scientific momentum of design, but coming down from the mountain were billowing storm clouds. Or were they smoke from new, powerful cannons hoisted up to the fortress walls? To this next wave of counterattack we now turn.

Beyond the Yale Conference

The War over Design Heats Up

Sound the Alarms!

The Intelligent Design Movement found solid encouragement in the Yale Design Conference in November of 2000, with its six packed evening sessions in the wood-paneled auditorium of Yale Law School, along with the oddity of a group of Darwinian protestors from Kansas who handed out anti-design literature to those who entered the auditorium for Phillip Johnson's kickoff address. I mentioned the conference's unexpected bonus—a breakout seminar with astronomer Guillermo Gonzalez, who argued that fine-tuning of the universe and planet Earth appears calibrated not just for life to flourish but also to provide ideal conditions for scientific discovery. Yet, whereas the symbolism of the Yale Design Conference was powerfully positive for theorists and friends of Intelligent Design, to ID critics this gathering was viewed as yet another sign of a steadily deteriorating situation.

Let me put a point on it. To Darwinists it was utterly appalling—it was virtually unthinkable—that after so much diligent educational effort on behalf of evolution at all levels of science education, the new ID arguments, especially those of Michael Behe, were being seized upon by tens of thousands of educated Americans, including hundreds of university professors, and were being turned against Darwinian orthodoxy like deadly weapons. This was deeply upsetting; it was *intellectual sabotage*, and it called for the most vigorous response. From the vantage point of those, like the late Carl Sagan, for whom the philosophy of materialism was a cornerstone of rationality,[1] this development was seen as one of the scariest breakdowns in human reason imaginable.

If there ever was a scientific idea worth defending, it was Charles Darwin's theory of evolution, a brilliant scientific breakthrough that in 2000, at the dawn of a new millennium, was "overwhelmingly established as fact."[2] True, it had upset many cultural apple carts when it exploded in Victorian England as the scientific bombshell of the nineteenth century. Nevertheless, said Darwinists everywhere, the important thing is that the theory had been vindicated over and over by accumulating evidence. Its power to enlighten mankind was viewed as incalculable. It had shaped, reorganized, and illuminated virtually every aspect of Western science and culture. It had eaten away many outmoded theories, ideas, and beliefs, thus functioning as the "universal acid" of our modern civilization, as philosopher Daniel Dennett had boasted.[3] According to its defenders, Darwinism was simply indispensable for a clear understanding of the world around us. Most famously, it had become the *central organizing principle of biology*. The late American geneticist Theodosius Dobzhansky was frequently quoted in this connection: "Nothing in biology makes sense except in the light of evolution."[4] Thus to discard Darwinian evolution is practically to hamstring biology; such a move would cast off the most important and secure knowledge that we have achieved!

A New Campaign and a Ubiquitous Slogan

Evolution was seen as the ultimate "hill to die for" within science, and thus opponents of design theory had been increasingly galvanized in the late 1990s about the attacks on their precious foundational knowledge, which they felt was undeniably secure. Yet the early books by Kenneth Miller, Robert Pennock, and Niles Eldredge and all the tireless efforts of Eugenie Scott and the NCSE office in Berkeley seemed not to have slowed the advance of ID significantly. So at the turn of the century, a new countercampaign gradually emerged. If the reader wants to slip inside the minds of leading Darwinists during this time to understand the perceptions, emotions, and rhetorical reactions going on, I recommend a quick reading of the introduction of Barbara Forrest and Paul Gross's book, *Creationism's Trojan Horse*. Yet let me issue one cautionary note: be prepared for a blast of bitter hostility and rampant distortion. Several readings of this chapter have netted, by my count, about fifty-two misrepresentations or outright errors in a space of merely nine pages of overwrought rhetoric.[5] What is most significant in this hostile torching of ID is the sense of danger, and the resultant dread, felt in the pit of the stomach by committed Darwinists. Something had to be done, and done quickly! By the middle of the decade, in 2004–2006, many facets of this countercampaign were revved up into high gear.

One of the most public acts in this countercampaign was that spokespersons for conventional science, often including entire national science organizations like the American Astronomical Society, began making announcements (usually by passing resolutions in executive councils) that declared that Intelligent Design is *religion, not science*. When I appeared on CNBC in a minidebate with anti-ID professor Brian Alters on February 24, 2006, he stressed the official rejection of ID by the National Academy of Sciences and the American Association for the Advancement of Science as decisive points. At times, such resolutions asserted that ID posed a threat to science education and to the future of America's place in the world as a leader in scientific research.[6] These sentiments were echoed in public declarations, verbally and in print, by Darwinian defenders, warning (again) that ID is *religion, not science*. This statement, forged like a rhetorical dagger to puncture and deflate ID's ballooning credibility, emerged as the number one talking point for ID opponents in this period. It was the most common, but the least nuanced, of the new sound bites against design.

This relentless negation of ID as science was beautifully captured in an article by attorney and freelance writer Dan Peterson, just as the vehement denunciations of ID had reached a furious peak in the fall of 2005. He set the stage by pointing out that in spite of "efforts of ID opponents to label them as 'creationists,' their arguments are not based on religious premises or Scriptural authority, and ID does not attempt to determine the identity of the designer. The inference that life is the product of an intelligent cause rather than unintelligent material forces may certainly have religious implications. But the arguments advanced by intelligent design theorists rely on neutral principles and facts drawn from mathematics, information theory, biochemistry, physics, astrophysics, and other disciplines." So, why would this new point of view upset the Darwinists and generate the rhetorical machine-gunning of ID? Peterson cleverly reviews the answer:

> If you ask ID's critics the reason for their opposition, they will tell you. Says the Dover teachers' union president, Sandy Bowser, "Intelligent design is not science." According to a caption in a *Washington Post* front page article, intelligent design is "not science." ID opponent and professor of physics and astronomy Lawrence Krauss goes on to explain that ID shouldn't be part

> of a curriculum because it's "not science." In a *Wired* magazine article that disparages ID, microbiologist Carl Woese contributes the point that intelligent design "is not science." Robert Pennock, a professor of philosophy who has been an active critic of intelligent design, elaborates that ID doesn't "fall within the purview of science." The lawyer suing the Dover school board contends that ID is "not science at all." The American Federation of Teachers adds helpfully that "intelligent design does not belong in the science classroom because it is not science." The National Science Teachers Association sheds a further bit of light, offering the view that "intelligent design is not science."[7]

Later I will explain how ID theorists sought to obliterate this incessant charge. What is germane here is that the "religion, not science" label, used sporadically in the 1990s, became a vehement first volley in almost every rhetorical encounter between the two sides. Often it was closely tied to the charge that many adherents of ID theory were known to be "religious" or, even more rhetorically poisonous, "fundamentalists."[8] The religious connections and affiliations of ID theorists and promoters was a central theme of the book by Forrest and Gross. Michael Behe began saying in his public lectures, "Darwinists say not to put stock in what I say because I have been seen entering and leaving churches." Of course Behe then reminds the audience that it was his Catholic teachers who kept assuring him that God had used evolution and that, on the contrary, it was the scientific evidence and arguments presented by Denton and others that led him to the design hypothesis.[9]

Another issue connected to the *science, not religion* put-down is much more philosophical: How is *science* to be defined? If scientists' research leads them to conclude that many features of organisms must have been designed by some intelligent agent, does that thought instantly become a religious conclusion? Have these scientists suddenly ejected themselves from the realm of science? To illustrate the clash over this issue, I'll relive a vivid moment on a local television program. I was invited in September 2005 to appear on *My Turn*, a news discussion program in Tampa, Florida, to discuss the explosion of ID. The three guests were asked by host Kathy Fountain to say what public schools should teach about origins. Eddie Tabash, an anti-ID attorney from California, repeatedly said: "ID should not be mentioned in schools because it is just theology; it's not science." After a break, I responded on air and said Tabash had it backward: ID scientists never *prejudge* in detecting design. *They never assume design; design must be positively detected*, by analyzing evidence and passing rigorous tests. Darwinism is different. It is profoundly theological in its basic operating rules, in that it lays down an *assured truth*—an axiom that amounts to a rigid religious catechism. It is this catechism then that serves as a starting point. The Darwinian catechism states that when scrutinizing complex living systems, one can rest assured that scientific evidence and logic can never lead one to conclude that there was an intelligent cause behind life. Thus the research begins with a rock-hard assumption that all aspects of nature are truly *undesigned*!

If I had been given more time on the program, I would have taken the argument one step further by pointing out the central dispute: How should we define "science"? Science has typically been defined—in line with the Darwinian "no design" rule—as the search for the "natural causes" of all phenomena (ID would substitute the phrase "real causes"). By this crucial philosophical decision, using the phrase "natural causes," the debate effectively has been decided before the first piece of evidence is laid on the table. In other words, the issue is simply settled at the level of the chosen definition of science. *Whoever wields the power to define science sets the terms that will decide the outcome of the dispute.*

Design theorists point out that the definition of science is itself not an issue that can be decided by the tools of science; rather, it is a tough question that philosophers have debated for many decades and continue to debate today. I wrote of this problem in some detail in the concluding chapter of *Doubts about Darwin*, and this crucial question is a main theme of Phillip Johnson's writings. Dembski wrote of the same problem in *Debating Design*, reminding the reader that "science may not, by a priori fiat, rule out logical possibilities. Evolutionary biology, by limiting itself exclusively to material mechanisms, has settled in advance the question of which biological explanations are true, apart from any consideration of the empirical evidence. This is armchair philosophy."[10] The fact that ID theory had rejected and discarded the biased *naturalistic* definition of science, which excludes the possibility of real design at the outset, was probably the most popular reason for branding ID with the stigma "religion, not science."

Denunciations, Books, and Fantasies

A second facet of the post-2000 campaign against ID was a series of statements and speeches given by educational leaders and public intellectuals (often in the form of published columns and editorials), which had the force of official denunciations and calls to arms. For example, in the fall of 2005 the presidents of Cornell University and the University of Idaho delivered public attacks on ID. Cornell president Hunter Rawlings III called on his faculty to help halt the movement's advance and begin reeducating America on such issues. Rawlings made no secret of his alarm at the spiraling state of affairs: *"This matter has become so urgent that I feel it imperative to make it the central subject of my State of the University Address."* Predictably, Rawlings's statement asserted that ID is "not valid as science." Drawing on a recently published attack by Allen Orr, he asserted that ID "has no ability to develop new knowledge through hypothesis testing, modification of the original theory based on experimental results, and renewed testing through more refined experiments that yield still more refinements and insights."[11]

Each of Rawlings's allegations was contested immediately by professors in the ID Movement, but just as the dust was settling from this bombardment, a second presidential pronouncement was published: the edict of University of Idaho president Timothy White. White simply tried to shut down all discussion of nonevolutionary views in science class: "At the University of Idaho, teaching of views that differ from evolution may occur in faculty-approved curricula in religion, sociology, philosophy, political science, or similar courses. However, teaching of such views is inappropriate in our life, earth, and physical science courses or curricula."

Because this is a rapid review, I must postpone the full reply of Intelligent Design to these charges. I will point out that almost every such public attack triggered a vigorous defense and counterattack from design theorists, generally posted within a week or two on the Discovery.org website, at ARN.org, or on one of the ID blogs. Public blasts from educators served mainly as dramatic symbols; they had relatively little persuasive value. In the case of the University of Idaho president's statement, it rang out as a *teaching-topic prohibition*. It largely backfired, sounding like a clumsy limitation of free speech. What was needed desperately in the view of Darwinists, more than sound bites and proclamations, was a point-by-point refutation of all major design arguments. Thus, one of the most important and concerted efforts made to halt the advance of ID after 2000 was the publishing of strong critiques of ID in scholarly journals and popular magazines. Also, new book-length critiques appeared with increasing frequency during 2001–2006. Besides the scientific content, these books and articles often included discussions of philosophical issues and cultural analysis of the design advocates themselves and their motivations. The arguments from these books will be encountered shortly. First, I need to set the scene of the struggle into which these rhetorical rockets flew, showing the astonishing power of fear-gripped imagination in the rhetorical onslaught against ID.

We must keep in mind that as ID advanced during the years after 2000, it produced a steady rise in the anger, contempt, and dread focused on design proponents and their arguments. In this process, strange fruit was borne. A scary mental landscape emerged, populated by nightmares, fearful predictions, and dire warnings. The authors of the two Oxford Press attacks on

ID (mentioned in chapter 1) led the way, indulging in an unprecedented rhetorical orgy of fear-fantasies. For example, Niall Shanks, a professor of philosophy at Eastern Tennessee State University, in his 2004 book, *God, the Devil, and Darwin*, whipped up hostility in the opening of his book by repeatedly employing his favorite epithet for creationists, "extremists." He then sought to rhetorically poison design theory by heaping the words *magic* and *supernatural science* on top of ID ideas and theory. His goal was clear: to thoroughly stamp design theory as dangerously extreme, utterly irrational, and thoroughly religious. In purveying this picture, there was no subtlety. Twice in Shanks's book, at the end of the preface and in the first three pages of his conclusion, the pages virtually explode with dozens of such hostile stigma-words.[12]

But ranging far beyond mere word-blasting of ID, Shanks paints the ultimate nightmare at the opening of his preface: "A culture war is currently being waged in the United States by *religious extremists who hope to turn the clock of science back to medieval times*. . . . [This] is an important fragment of a much larger rejection of the secular, rational, democratic ideals of the Enlightenment upon which the United States was founded. The chief weapon in this war is a version of creation science known as *Intelligent Design theory*." A few paragraphs later Shanks builds on his dark scenario and predicts where ID's "wedge strategy" is headed. He first envisions "intelligent design taught alongside the natural sciences." Next, a transformation of the educational system makes it more open to ID's "aim of religious instruction." Shanks's apocalypse then builds to its horrific climax: "At the fat end of the wedge, lurks the specter of a fundamentalist Christian theocracy."[13] In a nutshell, *fundamentalists will force religious instruction in public schools and then will seek to take over political control of the country!*

When I first heard of the *mere existence* of such an outlandish anti-design story line, painted as a picture of the goals of ID and of America's resultant danger, I found it hard to believe. Shanks's scenario seemed beyond absurd. I wondered: is he serious, or is he just indulging in apocalyptic fantasy to jar his followers? (Shanks has been severely criticized by scholars outside of ID. Philosopher Neil Manson commented on the book's "overblown rhetoric," adding that the "contempt Shanks displays is startling" and one particular "non sequitur is appalling." Del Ratzsch noted the repeated "cries that the sky is falling" and said the Oxford imprimatur is remarkable, given Shanks's heavy use of personal attacks on ID scholars. See the appendix for a key excerpt from Ratzsch's lengthy, biting review.[14])

Later I found that Shanks was not alone in this fantasy; others were repeating and embellishing the same nightmare. For example, the headline of an October 2005 web article by physicist Marshall Berman, posted by the American Physical Society, blares: "Intelligent Design: The New Creationism Threatens All of Science and Society." But the shrill title was just the tip of Berman's imagination-laden iceberg. He wrote of his own struggles against ID in New Mexico. After reviewing the progress made in halting ID encroachments in his state, he ratcheted up to a fiery conclusion, warning of the dangers of ID to science and "perhaps to secular democracy itself." He said design theorists were guilty of influencing politicians with little scientific background, and then he called for action: "Scientists must become more politically involved, if this assault is to be stopped. Replacing sound science and engineering with pseudo-science, polemics, blind faith, and wishful thinking won't save you when the curtain of 'Dark Ages II' begins to fall!"[15]

A similar fantasy appeared in the other Oxford book that appeared at the same time as Shanks's critique, and it was captured in the title—*Creationism's Trojan Horse: The Wedge of Intelligent Design*. Barbara Forrest and Paul Gross, the coauthors whose error-ridden and rhetorically frantic introduction I discussed above, wove the familiar scenario of an encircling group of fundamentalists, intent upon penetrating the fortress of academia, and ultimately plotting the replacement of liberal, rational democracy with a theocratic state. When my friend Joe, a retired high school English teacher, finished reading the first chapter of this book, he quipped, "Wow, Tom, perhaps with your connection with the Design Movement, you can land a cabinet position or an ambassadorship in the future theocracy."

The book's fantasies, which strongly echo what Shanks said, would seem barely to deserve mention, other than to further entertain the reader in the realm of the surreal. Yet note that these are the opening images in what is presented as a sober analysis of ID's history, its scientific claims, and its cultural goals. What's more, these multiple fear-fantasies all passed through a peer-review process at Oxford University Press. Astonishingly, the Oxford reviewers gave the green light to fantasy-themes of such shocking hyperbole, viciousness, and twisting of the truth that they might make political speechwriters blush. This hype of the dangers of ID, emanating from such sane and ordinarily credible scholars, confirms what Phillip Johnson began observing years ago: as Intelligent Design advances closer to its goals, many Darwinists react with "metaphysical panic."

The Wedge and the Paradigm Focus

Darwinian fears were often tied explicitly to certain stated goals that were related to Intelligent Design's oft-discussed *wedge analogy*. This was the basis of Shanks's fear-fantasy, and it also was the object of suspicion and harsh criticism by Forrest and Gross. In fact, the idea of a wedge formed the backbone of their critique, as seen in their subtitle, *The Wedge of Intelligent Design*. Nevertheless, since the time when it was developed by Phillip Johnson, design's main leader in the 1990s, it was never a secret concept. It even served as a prominent part of his 1999 book, which is named *The Wedge of Truth*. Seeming to have somehow forgotten this open and public aspect of the wedge, many critics of ID, with Barbara Forrest in the lead, have dramatized a leaked "Wedge Document" (posted on the Internet by ID critics starting in 1999) that had laid out goals for the next five years in a fund-raising campaign for the Discovery Institute. What was deemed sinister about this "leaked" document (which Discovery has affirmed more than once is indeed genuine) was a set of key points, two of which were said to be scandalous. This fed into the charge of ID's alleged plans for theocracy. Here I have placed the original Wedge excerpts within quotation marks, followed in each case by Discovery's published comment.

> "The proposition that human beings are created in the image of God is one of the bedrock principles on which Western civilization is built. Its influence can be detected in most, if not all, of the West's greatest achievements, including representative democracy, human rights, free enterprise, and progress in the arts and sciences."
>
> As a historical matter, this statement happens to be true. The idea that humans are created in the image of God has had powerful positive cultural consequences. Only a member of a group with a name like the "New Orleans Secular Humanist Association" [Barbara Forrest] could find anything objectionable here. (By the way, isn't it strange that a group [Discovery] supposedly promoting "theocracy" would praise "representative democracy" and "human rights"?)
>
> "Discovery Institute's Center . . . seeks nothing less than the overthrow of materialism and its cultural legacies." It wants to "reverse the stifling dominance of the materialist worldview, and to replace it with a science consonant with Christian and theistic convictions."
>
> We admit it: We want to end the abuse of science by Darwinists like Richard Dawkins and E. O. Wilson who try to use science to debunk religion, and we want to provide support for scientists and philosophers who think that real science is actually "consonant with . . . theistic convictions." *Please note, however: "Consonant with" means "in harmony with." It does not mean "same as." Recent developments in physics, cosmology, biochemistry, and related sciences may lead to a new harmony between science and religion. But that doesn't mean we think religion and science are the same thing. We don't.*[16]

To make sure that the charge of a "planned theocracy" was shown groundless, the Discovery Institute found it prudent to publish a document entitled "Discovery Institute and 'Theocracy.'" It mentions that the leadership and fellows of Discovery have been quite eclectic in worldview: Jewish, Roman Catholic, agnostic, and mainline Protestant. In its seven bullet points, the piece shows that the theocracy charge is baseless. Such attacks are "smears" and "ad hominem attacks" that "show the bankruptcy of the Darwinists' own position."[17]

Since the image and idea of an *ID wedge* became a key point of contention, we should note that several of Johnson's writings[18] use the wedge as a favorite metaphor. The wedge is a tool needed to break open a log that blocks the roadway to an understanding of scientific reality. *The log stands for Darwinism's grand claims and its materialist foundation*. How to remove it? By pounding a wedge into a crack and splitting the log so it can be removed. In this case, the crack is the separation between biological evidence on the one side, seen as powerfully pointing to a designer, and on the other side, the philosophical worldview of naturalism by which one *assumes* that no intelligence was ever involved in the origin of life or in the rise of its complexity and diversity. Johnson describes his role (public intellectual critic) as the "thin edge of the wedge"—helping to legitimize the discussion, and opening the way for younger scholars to follow.

The log in one sense could represent more than just a set of scientific theories or even its underlying philosophy (materialism or naturalism). It can be viewed as standing for the entire broad system of thought, practice, and well-developed doctrine that we have come to label "the Darwinian paradigm." In *Doubts about Darwin* I emphasized ID's rhetorical vision of dislodging Darwinism as the reigning paradigm, using Thomas Kuhn's notion of a paradigm in *The Structure of Scientific Revolutions*.[19] In Kuhn's book, a paradigm is a central concept, but it is not precisely defined. A paradigm includes (1) a set of research practices of a scientific community, rooted in (2) shared understandings and orientations of a great number of people in that community. Within a paradigm, strong arguments may surface, emotional clashes may erupt, but all members work together toward the same kinds of objectives, sharing key assumptions that guide research. In other words, a paradigm can be seen as a broad concept that is willingly *owned* by a network of researchers and that enables the network to be somewhat unified and at peace with itself about where it's going.

If Intelligent Design enters the picture and proceeds, as one of its goals, to split open the paradigm, to allow a newer, competing *design paradigm* to prove itself, then this constitutes a huge project indeed. (Of course, it is not an either/or situation; it is conceivable that a "competing paradigms" mode of science may persist for some time.) Critics tend to view this paradigm challenge as the ultimate ID pipe dream, and yet Michael Denton, the intellectual father of Intelligent Design, was clearly gesturing in this direction, referring to a Kuhnian "paradigm crisis" when he chose the title *Evolution: A Theory in Crisis*. His concluding chapter, "The Priority of the Paradigm," made a radical claim: *"There exists absolutely no evidence that biology's 'tree of life' actually sprouted its multitude of branches of different forms of life through the mutation-selection mechanism, and thus Darwinism inhabits the realm of scientific mythology. Rather, the only remaining glue that holds Darwinism together is the social power of the paradigm itself."*[20]

By the first decade of the twenty-first century, most members of the Design Movement saw Denton's thesis as being increasingly vindicated by each line of scientific and social evidence. They already viewed their work as laying the foundations of a new paradigm in biology. This paradigm talk appeared, for example, in the opening of William Dembski's 2004 book, *The Design Revolution*, which he described as a "handbook for replacing an outmoded paradigm (neo-Darwinism) with a new paradigm (Intelligent Design)."[21]

It is noteworthy that the total picture presented by design scientists and spokespersons was a radical repudiation of the "healthier than ever" picture of Darwinism presented by the media and in the textbooks used in schools and universities. To design theorists, the reigning paradigm was brittle and unresponsive to the cogent criticisms brought forward by ID. In effect, Darwinian orthodoxy was hunkering down and generally responding with a zero-concession strategy. This strategy boiled down to strict orders: *don't recognize anything of scientific merit whatsoever in the work of ID theorists*. This command captures the spirit of virtually all major ID-bashing books, both before and after the turning point of 2000, especially the pair of books from Oxford University Press discussed above, along with Mark Perakh's massive *Unintelligent Design* (which, oddly enough, attacks not only ID theorists but popularizers like Fred Heeren and even Bible-code advocates). The fourth anti-ID book, *Why Intelligent Design Fails*, is a collection of critiques edited by Matt Young and Taner Edis.[22] (Though taking a zero-concession stance like the others, the Young and Edis book is the most calm, careful, and respectful of the four books. It certainly deserves recognition for those qualities.) Many of the specific arguments of these books will appear in the chapters that follow, but their warlike essence has already been captured in this chapter. Lest the reader assume that the intensity of the debate over ID and Darwinism abated after the intense period of 2004–2005, I have included an "April Flurries" sidebar to give a sense of the intensity of scientific controversy in just one month: April 2006.

April Flurries

April 2006 witnessed a powerful flurry of articles, books, and blogged responses along the ID-Darwin frontlines. ID proponents hailed the release of *Traipsing into Evolution*,[23] a book critiquing the Dover decision, along with the tenth anniversary edition of *Darwin's Black Box*, with a powerful new afterword chapter. In the afterword, Michael Behe contends that his argument has grown significantly stronger through new discoveries, and he details why major arguments against irreducible complexity have failed.

Included in this swirl of news was a report on the discovery of a bizarre alligator-like fish called Tiktaalik Roseae[24]—claimed as a "spectacular link" between fish and land animals. It was called a "fishapod" because its characteristics seemed to help bridge the huge gap between fish and land-dwelling tetrapods. ID theorists pointed to its odd "mosaic" character—many fish or tetrapod characteristics seemed full-blown one way or the other rather than partially developed as if they were structures in transition.

Meanwhile, "after several years of claiming that there is no debate about the theory of intelligent design (ID)," three researchers published a study in the April 7, 2006, issue of *Science*,[25] claiming to have shown how an irreducibly complex system might have arisen through a few mutational changes. "This continues the venerable Darwinian tradition of making grandiose claims based on piddling results," says Behe. "There is nothing in the paper that an ID proponent would think was beyond random mutation and natural selection. In other words, it is a straw man." Behe explains that the authors are

> conveniently defining "irreducible complexity" way, way down. I certainly would not classify their system as anywhere near irreducibly complex (IC). The IC systems . . . contain multiple, active protein factors. Their "system," on the other hand, consists of just a single protein and its ligand. Although in nature the receptor and ligand are part of a larger system that does have a biological function, the piece of that larger system they pick out does not do anything by itself. In other words, the isolated components they work on are not irreducibly complex.

> Stephen Meyer adds, "If this is the best that the Darwinian establishment can do after ten years of trying to refute Behe's theory of intelligent design, then neo-Darwinian theory is in a world of hurt. Indeed, Behe's case grows stronger with each successive attempt to test it by experimental refutation."[26]

ID's New Crash Courses

The four books mentioned above that sought to shatter ID were hurled from the fortress in a concentrated barrage during 2004—somewhat like four earthshaking bunker-buster bombs suddenly pounding hardened underground targets. During the five years leading up to and including the crucial period of 2004–2005, design theorists were busy writing and publishing their own bunker-buster books. In fact, one major rhetorical trend that caused ID to *advance through heavy fire* was the sheer number of books that flowed from ID theorists. Not counting my sympathetic history, written originally as a Ph.D. dissertation in the rhetoric of science at the University of South Florida, some seventeen books and three educational videos were published in this brief period that were mostly or wholly positive toward design theory. Four of these twenty publications played a special role; they function as *crash courses*. These include, first, the peer-reviewed *Darwinism, Design, and Public Education*, from the Michigan State University Press, coedited by ID theorist Stephen Meyer and the rhetorician of science (and Darwin scholar) John A. Campbell.[27] This volume, a hefty paperback at over 540 pages, includes a surprising array of different points of view across the academic spectrum and serves as a crash course on when, how, and why public education can benefit from the insights provided by design theorists and their critiques of Darwinian theory.

Two more crash-course books were released in quick succession during 2004. The first, an eclectic mix, was from Cambridge University Press, *Debating Design*, edited by Michael Ruse and William Dembski. With its inclusion of twenty scholars from all major perspectives (pro and con), it was an instructive kaleidoscope of scholarly thought on the issue, and it helped legitimate the entire debate. The second is *The Design Revolution* by William Dembski. In forty-four short chapters it addressed virtually every major question (hostile or otherwise) tossed at ID in its brief history. Its importance and intellectual cogency as a response to Darwinists struck Ronald Numbers at the University of Wisconsin, the world's leading authority on the history of creationism. Numbers, not a theist, wrote the highly regarded and meticulously researched history, *The Creationists*.[28] He played a role in reviewing my own history of ID and recommending it for publication. With Dembski, he went a step further and provided a blurb: "For the past decade or so 'intelligent design' has stirred a storm of controversy. Is it nothing more than gussied-up creationism, as its critics charge, or a new scientific paradigm, as its advocates maintain? This sprightly catechism, written by the movement's leading theoretician, offers believers and skeptics alike (and I count myself among the latter) an authoritative, if one-sided, introduction to what the fuss is all about."

A final but momentous ID crash course was not a book or a conference but rather a pair of one-hour DVDs, *Unlocking the Mystery of Life* and *The Privileged Planet*.[29] These two documentaries, produced in 2002 and 2004, have the look of a *National Geographic* special. They were of such high quality that twenty-five PBS stations across the U.S. began showing the first one as it was made available. Both videos, produced by Illustra Media, are recognized for their cinematic craft and accuracy in presenting the ID side of the controversy. The second video grows out of the research of Guillermo Gonzalez and Jay Richards, which I discussed earlier, and appeared just months after their book, also entitled *The Privileged Planet*, was released in 2004.[30] The video documentary presents highlights of the book's arguments and evidence. Since the privileged planet hypothesis is relevant to the debate over fine-tuning, I will return to it in chapter 11.

The central thrust of the Intelligent Design Movement has been in biology, not astronomy or physics. Therefore the first video, *Unlocking the Mystery of Life* (hereafter referred to as *Unlocking*), is of overwhelming importance. Employing the rhetorical art of respectful understatement and the principle of "show, don't tell," *Unlocking* has won tens of thousands of allies for ID. Already by 2006, it had been translated into eleven foreign languages, including Russian, Chinese, Japanese, Spanish, German, Bulgarian, and Czech. In persuasive effect, *Unlocking* has come to overshadow all other ID crash courses or rhetorical tools. Using the hour of script time to maximum coverage, *Unlocking* resembles the pacing of a football game with its four-quarter division. One quarter, tracing Darwin's voyage in the 1830s and including shots of Darwin's finches on the Galapagos Islands, is a concise review of Darwinian theory. It emphasizes where natural selection works well (cyclical variations, or microevolution) and where it doesn't (the building of whole new body types and structures). The next quarter is an up-close overview of the work of Michael Behe and his concept of irreducible complexity—including the Darwinian responses to his book, *Darwin's Black Box*, and the ID replies to those criticisms. The third quarter focuses on the origin-of-life mystery and retells the story of Dean Kenyon, a biologist at San Francisco State University who in 1969 coauthored *Biochemical Predestination*, a pro-evolution book on the subject, and changed his mind a few years later. Special effects wizardry brings the viewer close-up with DNA coiled in the cell's nucleus and even shows the ribosome assembly-machine building proteins.[31] The final quarter showcases the work of Stephen Meyer on the information content (specified complexity) of DNA and also lays out Dembski's logic by which he is able to conclude that some particular object or system was in fact intelligently designed. In the five-minute conclusion, almost all the scientists who have been interviewed previously in the video return for a final comment. They explain why design is back on the table in science.

As I have interviewed many viewers of *Unlocking* over the past several years, I find a common pattern. The most often cited highlight is the special-effects 3-D moving image of the bacterial flagellum, with its forty protein parts, all of which are necessary for function. Michael Behe's work has attained a certain inescapable symbolism in such cellular nanomachines. Of course Behe, more than any other ID theorist, has attracted the heaviest attacks from critics of Design. Thus, it makes sense to turn now to Behe and his "black box of Darwin"—the remarkable living cell that the professor from Lehigh has helped us peer into with a new kind of vision.

5

Bombs and Rockets Galore

Michael Behe and Cellular Complexity

The central role of Michael Behe in the debate over Intelligent Design stems from his book, *Darwin's Black Box*, which exploded the scientific argument for design from the shadows of obscurity into the global limelight in 1996. His leadership since then has been focused on the defense of irreducible complexity and the ongoing task of developing and extending ID theory. In this labor, Behe has displayed a unique and colorful public persona. His writings on Intelligent Design (newspaper columns, journal articles, book chapters, and web essays) have projected a sharp intellect, expressing itself in a vivid and crisp writing style that is laced with generous doses of humor. Although the *quantity* of his writing has not matched Phillip Johnson or William Dembski, his *quality* of writing certainly has. The wit, clarity, and incisiveness of his prose has made Behe their equal, and this factor has done much to commend ID as a plausible stance, both to fellow academics and to the educated public.

Also vital to Behe's leadership of ID has been his willingness to interact—on the lecture stage and in print—with his most important critics. The content and style of his response can be summed up as politely aggressive. He has written detailed responses to many naysayers and has incorporated their criticisms into his slides in college lectures. His lecture that analyzes the strongest criticisms thus far says that critics have done ID a huge favor. His lecture's thesis (and the thesis of this chapter) is quite bold: *The irreducible complexity argument stands stronger than ever after the smoke of criticism has cleared; the withering attacks upon it have only vindicated it as a cogent argument for design.* Of course, the Darwinists vehemently deny this, as we will see shortly.

Behe's writings and lectures reflect just one side of his ethos (the perceived credibility of the writer or speaker) as the most visible ID scientist who is arguing publicly for design. His persona is engaging and complex—a blend of two strands. He is one part true-blue biochemist (his early work on Z-DNA won high marks among molecular geneticists).[1] He is also part "ordinary guy," typically dressed more like a deer hunter in Missouri than a biochemist in a high-tech university DNA lab. Lee Strobel, a Yale-trained legal journalist and author on ID, wrote about his encounter with Behe upon entering his office at Lehigh University:

> I knocked on the door of a nondescript office and was greeted cheerfully by Behe, dressed in blue jeans and a lumberjack shirt. He's enthusiastic, energetic, and engaging, with a quick smile and a crackling sense of humor. He always seems to be moving; even when perched on his swivel chair, he would roll back and forth ever so slightly. Wiry and balding, with wispy gray hair, a beard, and round glasses, he has a gentle and self-effacing manner that tends to put visitors at ease. Behe credits his casual manner to being the father of eight (at the time, going on nine) children, who keep him from taking himself too seriously. He laughed when I asked if he had any hobbies. "Mostly, I drive kids places," he said.[2]

With wife, Celeste, doing the homeschooling of quite a few of these nine offspring, the Behe household is indeed stretched, but according to Darwinist opponent Michael Ruse, it bodes well for the future of ID. He quipped, "Even if you can't beat us with your arguments, you're likely to take over by overpopulating us!"[3]

But enough of the generalities and fun stuff. In quoting Strobel above, I'm not trying to overemphasize the personal side, and he isn't either. In fact, much of his excellent chapter on Behe relates the academic sniper fire that has been triangulated upon his arguments. The defenders of Darwinism's fortress of public credibility have been more concerned about Michael Behe than any other ID biologist—*by far*. Critic Niall Shanks, in his chapter "The Biochemical Case for Intelligent Design," places Behe in his crosshairs. At the outset of his attack, he recognizes that

> *Darwin's Black Box: The Biochemical Challenge to Evolution* (1996) . . . is without doubt the most influential of the recent books written in support of intelligent design. On its face, it is an attempt to articulate a principled argument from the study of nature to the conclusion that nature contains features that require intelligent design. The argument is derived from a long

> line of arguments . . . that culminated in Paley's version of the design argument. *At the thin end of the wedge, much of the intelligent design movement can be thought of as footnotes to Behe.*[4]

Shanks later adds, "Behe's arguments are the linchpin of the intelligent design movement."[5]

Behe's work is a serious threat among Darwinists, but it is virtually employed by the Discovery Institute as the basis for defining ID: Intelligent Design is the theory that "holds that certain features of the universe and of living things are best explained by an intelligent cause, not an undirected process such as natural selection."[6] "Features of the universe" may invoke the fine-tuning arguments discussed in Guillermo Gonzalez and Jay Richards's book, *The Privileged Planet*, but with ID's emphasis on biology, the phrase "features . . . of living things" certainly triggers the image of irreducibly complex machines as much as it does anything else.

So Behe's centrality to ID is acknowledged by all sides, and the level of media focus on his work has been massive. I phrase it this way: the waters of sophisticated skepticism were building up for many years, but when Behe published *Darwin's Black Box* in 1996, the dam broke. We've discussed Behe several times in the opening chapters, so his theory only needs a brief review here. However, because of our theme—Darwin "striking back" through his dogged latter-day defenders—we start that review by casting our first glance at "Darwin's wager," which hovers at the core of the controversy. Then, we will tour three battlefields, noting both the rockets of criticism and Behe's defensive shields (assisted by Dembski). Finally we will descend to the innards of cellular complexity and probe into the preprogrammed architecture of individual proteins. At this level, we confront a widely overlooked yet still-aching mystery that makes the case for irreducible complexity far more powerful than most commentators are aware of.

Back to the Beginning: Darwin's Wager

One frustration among Darwinian defenders who face Behe is his use of a crucial test proposed by Darwin himself in the *Origin* (now famous through Behe and others).[7] Darwin says: "If it could be demonstrated that any complex organ existed which could not possibly have been formed by numerous, successive, slight modifications, my theory would absolutely break down."[8] Darwin writes this in his sixth chapter, "Difficulties on Theory," while addressing the problem he labels "Organs of Extreme Perfection and Complication." Even though Darwin may seem to have been making a friendly gentleman's wager, or perhaps issuing a *confident prediction* from his theory (and a way to falsify it), to me he seems to have used the quote more as a clever rhetorical device. He is referring to what he considers an extremely remote possibility, which the evidence of his day fell short of. In fact, he asserts right after this famous quote that he "can find out no such case" where an organ cannot be formed by tiny successive changes, using his imagination and the data available to him. This entire section of the chapter on difficulties is filled with speculations about "intermediate or transitional grades" in forming, for example, eyes and lungs and electric organs of fish from earlier and more primitive structures.

While Darwin sensed the seriousness of such problems, he argued that his theory could in fact extricate the doubting reader from his doubts. Natural selection can be imagined building complex organs if you just observe in nature intermediate-grade organs—such as a variety of simpler eyes, or any of the other series of intermediate grades mentioned in this chapter. Just four pages after the break down quote, Darwin comments on this whole matter of small-step development of new complex structures. It is a vital part of his argument. I beg the reader's patience as we enter this tiny sliver of the *Origin*, so as to feel the full force of his appeal. He argues for the superiority of viewing natural selection as the agent that crafted these organs, and not a creator:

> Although in many cases it is most difficult to conjecture by what transitions an organ could have arrived at its present state; yet, considering that the proportion of living and known forms to the extinct and unknown is very small, I have been astonished how rarely an organ can be named, towards which no transitional grade is known to lead. The truth of this remark is indeed shown by that old canon in natural history of "Natura non facit saltum." [Nature does not make a leap.] We meet with this admission in the writings of almost every experienced naturalist; . . . nature is prodigal in variety, but niggard in innovation. Why, on the theory of Creation, should this be so? Why should all the parts and organs of many independent beings, each supposed to have been separately created for its proper place in nature, be so invariably linked together by graduated steps? Why should not Nature have taken a leap from structure to structure? On the theory of natural selection, we can clearly understand why she should not; for natural selection can act only by taking advantage of slight successive variations; she can never take a leap, but must advance by the shortest and slowest steps.[9]

Thus Darwin asserts his theory is plausible after all, even in the teeth of a would-be objector who said, "How did the eye evolve?" Darwin in effect shot back, "If creation were the correct theory, then why would the creator produce so many intermediate forms of simpler eyes in the animal kingdom, leading all the way down to a light-sensitive spot? Is he trying to mislead us? Does not this evidence suggest powerfully that natural selection, ceaselessly active in recruiting new organic parts, to the constant betterment of a specimen in its fight to survive, has knit together ever more and more sophisticated eyes as you ascend the tree of life?"

Putting words in Darwin's mouth can be a tricky endeavor, but I think I've accurately expressed both his sentiment (subtle pathos), and the core of his logos, or argumentative thrust, seen in this famous chapter. I can sympathize with Darwin a bit. He knew that he had a powerful idea in his hands—a simple concept that seemed to have unlimited potential to explain everything we see in the world of nature. Yet he also knew that complex organs, like the eye, were (to most intelligent observers) a hard-to-explain aspect of nature on his theory, and he was doing his rhetorical level best to show how natural selection can be *imagined* (at least) to mimic the work of a creator.

A quick personal flashback shows that contemporary Darwinists reason as Darwin did. I once asked the renowned Princeton biologist John Tyler Bonner how he would explain the macroevolution of such complex organs. In response, he directed me to George Gaylord Simpson's book, *The Meaning of Evolution*. Simpson's answer—looking at a suggestive sequence or gradation of different eyes, from simple to complex—was not much advanced on Darwin's. Do we really know that natural selection can accomplish the drastic morphological transitions between these different eye types, with all the knitting and organizing of new complex proteins? I was not at all impressed with the Bonner-Simpson "Just So Story."

Pressing the Wager into Service

Talking about the imagined gradual development of complexity at the level of an organ is one thing; looking at such step-by-tiny-step development inside the cell, at the level of molecules, is quite another. In a sense, it is easy for human imaginations to step along the imaginative path from a simple light-sensitive spot, then to a "cupped" primitive eye, then to a simple fixed eye with lens, and finally all the way to a vertebrate eye. The human mind has a well-known ability to compare similar images and patterns and to mentally morph across the gap from one to another, along a purported pathway of evolution. (Computer image-morphing programs now give visible expression to this long-standing human ability.) My own theory is that this morphing capacity of the human mind stands behind much of the confidence in such Darwinian macroevolutionary scenarios as seen in the eye sequence of Darwin and Simpson. But what about morphing along a simple-to-complex pathway on the way to evolving molecular contraptions inside the cell? Can these nano-organs also be plausibly envisioned as being "formed by numerous, successive, slight modifications"?[10]

It is right here that Behe carved out a new testing ground for Darwin's wager. He stated his intent, in using Darwin's test, to focus on the Lilliputian world of cellular systems comprised of interacting proteins. In *Darwin's Black Box*, Behe spreads out in horrendously complex detail the results of decades of research by molecular biologists. These tiny molecular machine parts that work together in marvelously integrated concert are traced in each of the seven systems chosen. (Behe says that such systems swarm throughout the cell.) One comment heard frequently, even in hostile reviews, is that Behe does an impressive job of describing the sheer complexity of such multipart (multiprotein) systems. At the start of a lengthy attack on Behe in *Unintelligent Design*, Mark Perakh says, "All these systems [blood clotting, the cilium, etc.] look like real miracles and it is fun to read Behe's well-written discussion of those immensely complex combinations of proteins, each performing a specific function. The complexity of the biochemical systems has been demonstrated by Behe in a spectacular way."[11]

This is where Behe takes up Darwin's test (can X be formed plausibly by numerous, successive, slight modifications?) and applies it to such cellular machines and systems. Darwin's theory, says Behe, must submit to its ultimate test here in the molecular contraptions. The results of the test are now in, and natural selection has *flunked spectacularly* as a credible explanation of the protein assemblages we find chugging along within a cell's daily routines. Because of their reliance upon a specific mandatory set of proteins in order to retain function, these systems are irreducibly complex. Take one protein away and they can't function any longer. The notion that they were evolved, step-by-molecular-step, seems overwhelmingly implausible, at least in terms of natural selection, which cannot look ahead and select for a faraway goal. These irreducibly complex entities apparently were designed, not evolved.

Of course, Darwinists will allege that some intermediate forms of these systems do exist, at least in two cases—the flagellum and the blood-clotting cascade. We will turn to those two cases soon, but remember Darwin's question: "Why should all the parts and organs of many independent beings, each supposed to have been separately created for its proper place in nature, be so invariably linked together by graduated steps?"[12] Unfortunately for Darwin, the "graduated steps" that have been tracked down in nature for some organs simply do not exist in the cell. There are no graduated series of simpler forms of molecular machines. Furthermore, says Behe, if you search the literature of evolution and molecular biology for detailed, testable scenarios as to how any of these systems came about, you find what he calls a "thundering silence." No one in the world of science has a clue as to how these could be evolved, step-by-molecular-step. This final bit of research by Behe acts as the clincher of his argument.

That, of course, is Behe's core argument, although he develops it in much more detail in his book. He includes side discussions of the origin-of-life mystery (which we will confront in chapters 8 and 9), and he probes the self-organization ideas of Stuart Kauffman. It is obvious that *Darwin's Black Box* should be mandatory reading for both friend and foe of ID, to enable one to grasp the theory's essence and to understand why ID has been compelling to many educated Americans not philosophically committed to naturalism, who have taken time to read Behe's book.

Darwinist critics have unleashed successive attacks upon Behe's arguments. Some took the form of hostile book reviews—there have been well over one hundred book reviews published since 1996; about half were somewhat hostile or worse. Other critiques appeared as journal articles and book chapters, as in the four bunker-buster books dropped on ID in 2004. These counterarguments boil down to a half-dozen major types, three of which I have room to sketch out below, along with Behe's rejoinders. I should point out that Behe has been vigorous in response and increasingly optimistic in the face of this heavy criticism. He knew his arguments would draw very heavy fire, and in his responses he has repeatedly pointed out that Darwinian researchers are still at a complete loss—in terms of testable scenarios—as to how irreducibly complex systems arose.[13]

Flawed Analogies?

One major approach in blocking the conclusion of design from the irreducible complexity (IC) of machines inside the cell is to attack Behe's mousetrap analogy as "bad" (or "flawed"), while providing in its place a "good" Darwinian analogy that promises to show how IC can indeed arise from nature. I won't waste much time on this area, since analogies are just that—tools of illustration. The argumentative force on either side, for or against the IC argument, doesn't ultimately rest on the power of one's physical parallels. Yet the discussion is prominent enough for us to look into the "flawed analogy" accusation.

When the *Firing Line* debate took place in December 1997 at Seton Hall University, which was shown on PBS stations across the U.S., it featured four participants on each side of the ID question. One of the most interesting pairings was a ten-minute conversation between Behe and Kenneth Miller.

You'll recall Miller's book, *Finding Darwin's God*, one of the most prominent anti-ID works mentioned in chapter 3. One of Miller's statements was that the mousetrap analogy failed, thus Behe's argument is hobbled. Just remove the catch, twist the holding bar a bit, and its tip can be carefully perched under the spring, ready to be sprung. Miller even demonstrated the snapping of his modified four-part mousetrap (to the befuddlement of moderator Michael Kinsley, who asked for an explanation of what was going on). Miller says that because Behe's illustration fails (the trap can work with less than five parts) then his theory is brought into question.[14]

Behe immediately rejected Miller's argument, pointing out that the modification required intelligent monkeying with parts, with the spring now serving a new, second function as the catch. Behe mentioned another assertion he had heard—that the mousetrap failed because you could eliminate the base and attach the other four items to a floor. This absurd disproof was dispensed of quickly, as Behe noted that the floor is now serving as the base. Five parts *are still needed*.[15]

Such mousetrap criticism has proliferated ever since this debate. More vigorous along this line was biologist John McDonald at the University of Delaware. McDonald is known for his webpage that posts various simpler mousetraps: one-part (he admits it doesn't work well), two-part, three-part, and four-part.[16] After overhauling his mousetraps in recent years (in response to criticism from ID theorists), he reposted his cartoonlike figures so that they now pass through thirteen stages of evolution, from one part made of a wire loop, all the way to the modern snap trap. Nevertheless, they still cannot smoothly transition from one to another without significant monkeying and reworking. Also, McDonald is well aware that the biological pieces in real living cells, for which the mousetrap parts serve as pictures, are specially shaped proteins that are dependent upon a careful digital blueprint in the DNA. There is no reference in McDonald's mousetrap argument to the fact that proteins are uniquely shaped and highly improbable pieces. Proteins aren't like pieces of wire or lumps of clay. They can't be slightly modified as needed by bending, twisting, and molding. In summary, to anyone who is not philosophically wedded to the Darwinian answer, McDonald's evolving traps seem an exercise in irrelevance.

Nevertheless, in Kenneth Miller's webpage critiquing Behe's mousetrap, he asserts that the fact that intelligent intervention is needed to go from one mousetrap to another misses the point. The problem for Behe, says Miller, is that you *can catch mice* (albeit inefficiently) with fewer than five parts. Also, the mousetrap precursors can be used for other purposes. Miller sometimes sports a tie clip made up of a tiny base, and an even tinier spring and hammer. He says, "In showing that it is possible to use part of a mousetrap for a different purpose, one shows by analogy that it is also possible to use part of a biochemical system for a different purpose. That's the fatal danger of the mousetrap analogy for Behe's argument, and it has become a trap from which he cannot escape."[17]

Behe's response is that there is no fatal danger to escape from. He says he has never denied that subsystems of a machine that has irreducible complexity *could possibly serve another function*. The problem is in what drives the transition from one system to another—a plausible, testable scenario to move from the simpler system to the more complex one. The whole idea of *functional precursors* has the idea of contraptions working for the same function, or even different functions, *along a series of step-by-step developments that make evolutionary sense as driven by natural selection*. That is what has never been demonstrated, says Behe, in the literature critiquing ID or the mousetrap in the years since the publication of *Darwin's Black Box*.

In *Debating Design*, the Cambridge Press forum on ID released in 2004, Behe and Miller square off once again, each being given a chapter. Behe says: "But that is exactly what it [McDonald's series] doesn't show—if by 'precursor' Miller means 'Darwinian precursor.' On the contrary, McDonald's mousetrap series shows that even if one does find a simpler system to perform some function, that gives one no reason to think that a more complex system performing the same function could be produced by a Darwinian process starting with the simpler system. Rather, the difficulty in doing so for a simple mousetrap gives us compelling reason to think it cannot be done for complex molecular machines."[18] Meanwhile, other scholars such as William Lane Craig question the relevance of this line of attack altogether, saying that it is entirely beside the point in evaluating IC as an empirical pointer to design.[19] Some Darwinists have countered with analogies of their own (such as stone arches) to fight Behe's analogy.[20]

Course Extensions and Gestures toward the Unknown

A second major way that Darwinists attack is in the area of *time extensions* and *scientific imagination*. Earlier I said that Darwinism receives a resounding "F" from Behe as a plausible explanation of molecular machines that manifest IC within the cell. Many Darwinists protest this grade vigorously. They argue that the course grade for natural selection, when it comes to cellular complexity, is not "F" but rather "I" for "Incomplete." In effect, the Darwinist says, "We just need an extension, if you don't mind. Science may not yet have a detailed explanation for the molecular systems Behe describes, but there is every reason to think that in the decades to come we will find the pathways by which such systems were assembled. To call it quits on this research at this point is extremely premature."[21] This argument takes several forms, one of which is the charge: *"Intelligent Design is just an exercise in personal incredulity. They personally can't see how it could happen, so they throw up their hands, say they can't see how it could happen, and just give up."* In fact, the repeated response to Behe from the renowned Oxford biologist Richard Dawkins is along this line. Dawkins accuses Behe of simply being lazy; he should get into the research lab and work hard to find those pathways by which the cells' irreducibly complex equipment was built.[22]

The argument that ID theorists personally cannot imagine how molecular machines arose, and therefore it is just a failure of their own scientific imagination, prompted this tart response from William Dembski:

> Miller [in an earlier form of his chapter on the flagellum in *Debating Design*] claims that the problem with anti-evolutionists like Michael Behe and me is a failure of imagination—that we personally cannot "imagine how evolutionary mechanisms might have produced a certain species, organ, or structure." He then emphasizes that such claims are "personal," merely pointing up the limitations of those who make them. Let's get real. The problem is not that we in the intelligent design community . . . just can't imagine how those systems arose. The problem is that Ken Miller and the entire biological community haven't figured out how those systems arose. It's not a question of personal incredulity but of global disciplinary failure (the discipline here being biology) and gross theoretical inadequacy (the theory here being Darwin's).[23]

As a matter of fact, Behe—busy with his own major research projects on DNA—has welcomed any such continued research on the pros and cons of IC theory, attempting to falsify his main argument by providing scenarios that are testable (and which pass the test) concerning how subcellular motors and systems could have evolved. And indeed, the process of research into such systems is still rolling along, as we will see.

Sometimes the "give us more time" plea is linked with an assertion that there are so many interesting ways of generating novel sequences and configurations in biological systems that perhaps one of these new and very general ideas might pan out. ID critic Mark Perakh tried to derail Behe by offering four such "novel approach" scenarios, all of which struck me as exercises in vigorous hand-waving. (Note: We are entering slightly more difficult conceptual terrain. If you get bogged down, just skim the remaining part of this section.) For example, Perakh supposes that a typical blood-clotting protein, which may have one hundred amino acids, would have a probability of one chance in 8.03×10^{59} (8 followed by 59 zeros) of forming. This, he admits, is too remote. However, he theorizes that perhaps the one hundred amino acids can be seen as arrangements of ten "bricks" with ten amino acids each, and the group bricks of subunits get shuffled as bigger units. Now the chances are much better: there are only 534,800 ways to shuffle the bricks. Because the chances are much better, says Perakh, the protein might be built by chance! But there are several problems, none of which is even mentioned by Perakh. For example, where is the evidence (1) that amino acids are ever arranged in such movable bricks, (2) that there is a mechanism to shuffle the bricks promiscuously at a predictably high rate, or (3) that such brick-by-brick arrangements are any more likely (on a unit basis) to serve as ideally shaped proteins than the surrounding vast ocean of sequences that fall outside the brick patterns.

Of course there is no evidence for these three points. Perakh's idea falls flat. His other three hypothetical ideas are similarly lacking in plausibility as pathways on the road to useful protein sequences, let alone to an entire multiprotein system.

At the end of this discussion, Perakh says, "Even if options (1) through (4) can be ruled out somehow, what about some mechanisms (5), (6), etc., that we didn't think of yet? To assume that everything in nature happens only according to known mechanisms would unduly limit the path to the scientific elucidation of the unknown."[24] In other words, we are told that there are many interesting, new theoretical approaches, and since there may be an indefinitely large number of other research approaches or ideas that we haven't even thought of yet, why not give it a try and use research money to explore these avenues?

Superficially, the appeal rings plausible. Unfortunately, when the details of Perakh's ideas are examined, they severely disappoint. He seems to sense this himself and immediately adds this side comment:

> I can foresee a common . . . counterargument to the notions presented above—an assertion that these scenarios are "just so stories" that do not prove anything, because there is no direct empirical evidence of their actual occurrence. *Indeed all these scenarios are speculative.* They prove, though, one thing—that the claims of creationists asserting the alleged impossibility of evolution because of the extremely small probability of its individual steps are invalid insofar as they consider only the exceedingly unlikely combination of chance events as if such purely random chains of events were the only option. *In fact, nature has in its arsenal a host of other options ignored by the creationist scenario.* Unless all these options are shown to be impossible (which is not the case by a long shot) the creationist claims remain much more speculative than those listed above or than many other "natural" scenarios we have so far not even imagined.[25]

There are several ways to reply to this enthusiastic "Nature can do it after all!" type of answer. One is to point out that the supposed "arsenal" filled with "a host of explanatory options" is a vague promise, not a solid, itemized stockpile. *His wording here is sheer empty rhetoric.* Where does Perakh present even a modest list of empirically rooted potential explanations of the origin of complex machinery? Nowhere in his chapter, nowhere in his book, nor for that matter in any other book that I am aware of.

A second reply is simply to point out (following William Dembski) that design theory engages in "eliminative induction"—a phrase that can be compared with Sherlock Holmes's "famous dictum: When you have eliminated the impossible, whatever remains, however improbable, must be the truth."[26] That is, any feature in the world can be explained by scientific law (necessity), by chance, or by design. There are no other options. Explanatory elements can be mixed, with law and chance working together (like a hurricane, whose dynamics obey physical laws but whose path is unpredictable due to chance factors). Even when design is used as an explanation for how "Tom loves Normandy" appears written in the sand at Clearwater beach, the sand particles are obeying the laws of physics and chemistry, but those laws will not suffice to explain the message by themselves. So in design theory, if law, or chance, or even their combination appear by our analysis to be overwhelmingly implausible as explanations, we go with the one remaining *good* induction: it was intelligently designed.

Yet Perakh's plea was for us to place hope in mechanisms that we didn't think of yet or in his "explanatory arsenal"—which seems nonexistent because it is promissory at best. Look again very carefully at his words: "Unless all these options are shown to be impossible (which is not the case by a long shot) the creationist [his stigma-word to tar ID] claims remain much more speculative than those listed above or than many other 'natural' scenarios we have so far not even imagined." How can Perakh say that ID's design inference is "much more speculative" than vague, gestured-at, currently unimagined hypotheses involving chance and law? Unless he can convince us that he possesses a "private metaphysical pipeline to ultimate reality"[27] that assures him with 100 percent reliability that his worldview of naturalism is correct and that no mind exists apart from matter, *how can he rate the design inference worse on the speculative scale than his own speculated (but unknown) mechanisms?* It seems that a preferred brand of metaphysics has quietly arisen here and asserted itself. A promising explanatory idea (design) was simply *vetoed* by Perakh's preferred philosophy.

Dembski's statement on this Sherlock Holmes–type eliminative task makes an added point: the presence of a "can-do premise." This idea says, "*We all know quite well the structure of the physical universe, especially its cause-effect structure, and certain kinds of complexity are known to arise easily from intelligence, and these same kinds of complexity are never seen arising from brute nature."* Keep in mind that "specified complexity" here is the umbrella term for several kinds of complexity (including that in DNA and proteins), and Behe's irreducible complexity fits under this umbrella. Dembski says,

> That's why intelligent agency having the causal power to produce systems that exhibit specified complexity is such an important premise in eliminative inductions that attempt to infer biological design. Let's even give this premise a name: *the can-do premise* (because we know that designers "can do" it, that is, they can generate specified complexity). Precisely because intelligent agency is reliably correlated with specified complexity, there is no need to give equal weight to every conceivable naturalistic hypothesis or to wade interminably through the never-ending list of half-baked, handwaving Darwinian just-so stories, none of which has ever given any evidence of actually elucidating biological systems that exhibit specified complexity. In other words, the can-do premise makes the eliminative induction here a local induction that can legitimately infer design.[28]

Dembski admits in the next paragraph that "eliminative inductions, like all inductions and indeed all scientific claims, are fallible." But that does not for a second suggest we toss out the powerful role of such inductions. He says they "need a place in science. To refuse them, as evolutionary biology tacitly does by rejecting specified complexity as a criterion for detecting design, does not keep science safe from disreputable influences but instead undermines scientific inquiry itself."[29]

The Flagellum and Blood Clotting: Halfway Houses?

The third major way of fighting the argument from irreducible complexity considered here is seen in the most famous of the IC battlefields. Here the conflict is over certain irreducibly complex systems that Kenneth Miller asserts loudly have been evolved, or certainly can be evolved. Behe has published articles and book chapters that argue just as loudly that Miller's arguments themselves collapse upon close inspection. One ironic aspect of Miller's claim to have falsified Behe's argument over and over is that he has said repeatedly that Intelligent Design theory is not science; it is not falsifiable. Yet in the next breath, he claims to have falsified Behe's theory of irreducible complexity! Another common charge against ID is that it makes no predictions. This is fallacious. The important point here is that Behe's entire argument is essentially a clear and daring prediction: *Darwinists will not begin filling in plausible, testable scenarios for any of the irreducibly complex cellular systems.* That's where the most intense tussle is ongoing.

One argument that we don't have time to go into involves the work of a biologist named Barry Hall, who managed to replace one genetic unit, the "lac operon," with another unit that already existed elsewhere on the genome of the bacteria. Miller announced that this system possesses IC and has been evolved right in the lab. Here is a snippet from Behe's article replying to Miller. Don't worry if you don't get the details because you haven't read the preceding article. But it gives a clear sense of the type of intense interaction going on. Behe says:

> Miller ends the section in his typical emphatic style: "No doubt about it—the evolution of biochemical systems, even complex multipart ones, is explicable in terms of evolution. Behe is wrong." (Miller, *Finding Darwin's God*, p. 147)
>
> I disagree. Leaving aside the still-murky area of adaptive mutation, the admirably-careful work of Hall involved a series of micromutations stitched together by intelligent intervention. He showed that the activity of a deleted enzyme could be replaced only by mutations to a second, homologous protein with a nearly-identical active site; and only if the second repressor already bound lactose; and only if the system were also artificially supported by inclusion of IPTG; and only if the system were also allowed to use a preexisting permease. Such results are exactly what one expects of irreducible complexity requiring intelligent intervention, and of limited capabilities for Darwinian processes.[30]

This same level of intense interaction pervades virtually all of the exchanges between Behe and Miller and between Behe and another nemesis, the pioneer in blood-clotting research, Russell Doolittle. Space allows us two examples of such clash: blood clotting, which is summarized tightly, and the flagellum, which I'll expand on.

The setting up of a blood clot is a thing of complexity and wonder. It seems inordinately complicated, since a sequence of twenty different proteins trigger one another like dominoes falling in order, until the fibrin mesh is finally set up for the clot itself. Behe argues that this irreducibly complex cascade could not evolve, since you don't get a clot unless all twenty are present. It just couldn't have evolved step-by-step. Of Darwinian responses to Behe on this point, the most famous was an article by Doolittle, whose excellent biochemical sleuthing was a major part of the unraveling of this clotting pathway. In an article in *Boston Review* in 1997, Doolittle said Behe was wrong about the cascade. It wasn't irreducibly complex. A researcher named Bugge had "knocked out" one blood-clotting protein, prothrombin, and the mice suffered because clots couldn't be cleared. They knocked out another protein, fibrin, in a different group, and the mice suffered because clots wouldn't form in the first place. Then, in a third experiment, they knocked out both, said Doolittle, by interbreeding the two deficient lines: "And what do you think happened when these two lines of mice were crossed? For all practical purposes, the mice lacking both genes were normal. Contrary to claims about irreducible complexity, the entire ensemble of proteins is not needed. Music and harmony can arise from a smaller orchestra. No one doubts that mice deprived of these two genes would be compromised in the wild, but the mere fact that they appear normal in the laboratory setting is a striking example of the point and counterpoint, step-by-step scenario in reverse!"[31] Behe, when discussing criticism of IC theory, points out that the massive problems of the mice with both proteins missing is the same for mice with just fibrin missing—*clots cannot form at all!* So the "smaller orchestra" is hamstrung; Behe seems thoroughly vindicated. As a result, the whole point of Doolittle in citing the study of Bugge backfires.

The last attack on Behe is the most contentious battlefield of all: the bacterial flagellum. This microscopic outboard motor, sometimes called the "mascot of ID," became famous through the *Unlocking the Mystery of Life* documentary (hereafter referred to as *Unlocking*), where almost fifteen minutes of discussion of Behe and irreducible complexity are dominated by discussion of the flagellum, aided by dazzling computer graphics of this organelle twirling at high speed. Scientists on camera explain the difficulty of evolving a tiny rotary engine with bushings, propeller, driveshaft, U-joint, and several kinds of ring structures. *Unlocking* adds two bonus points:

1. A flagellum researcher, University of Idaho biologist Scott Minnich, points out another level of complexity not captured by Behe's argument (Behe focused mainly on structural proteins). Minnich, a leading ID biologist, says that many flagellar proteins control the *construction process*, switching the building phases on and off at just the right time and setting up construction in the proper sequence. The complexity of this process equals or exceeds the physical complexity of the flagellum itself, and the plausibility of the step-by-step development of this set of controls is vanishingly small.
2. The common counterargument, "co-option" (also called "co-optation") is explained. In the case of the flagellum, with forty proteins, this argument points out that at the flagellar core is a needle-nose injector pump—an assembly of ten proteins that sometimes appears alone, penetrating the cell wall. In certain bacteria (e.g., the bubonic plague bacteria of the Black Death fame), this subassembly injects toxins into

> host cells. This injector pump, called the "Type Three Secretory System" (TTSS), is said to have evolved first, and then it evolved as it "co-opted" or borrowed other proteins to achieve a new use. Eventually, after thirty other protein parts were added, the flagellum had evolved.

In the *Unlocking* video, Scott Minnich stands in his microbiology lab and quietly assesses the Darwinian TTSS scenario. Yes, he says, it is remotely possible that the TTSS injector came first, and he affirms that its ten proteins do seem to parallel or match the core proteins of the flagellum. But that's where you bump into a huge problem. Where did the cell find the other thirty or so proteins to build incrementally from the TTSS all the way to a rotary-motor flagellum? You come to the point where you are borrowing from nothing, and the plausibility of the scenario fades quickly.

Since *Unlocking* was filmed (in 2000–2001), Minnich has done extensive research and has published in a peer-reviewed journal[32] his findings showing that *the flagellum is likely to have historically preceded the TTSS.* This is indicated since the TTSS is found in *gram negative bacteria* that seem to have appeared in a later era, when more advanced kind of cells called *eukaryotes* had appeared. These gram negative bacteria with TTSS injectors don't hassle other *prokaryotes*—bacterial life-forms. In essence, the current best evidence indicates that a flagellum devolved (decayed) into a tiny subsystem, the TTSS injector pump.

Many observers watching the shifting battles over Behe's theory feel that Kenneth Miller was premature in loudly declaring victory, insisting that the flagellum could plausibly have evolved from the TTSS, when the evidence indicates that the TTSS was the fruit of reverse-evolution. Miller's exercise in hand-waving (arguing that the TTSS led right on to the flagellum) has always depended upon the other thirty proteins—floating in from the cellular environment. But what's the source? Are they just easily bubbling up from day-to-day cellular processes, in wondrous variety, ready to be recruited to build from ten TTSS proteins up to the flagellum's set of forty? It would be wise here to look closer at proteins and confront their mysterious origin.

A good summary of the problems with the TTSS hypothesis was brought out by William Dembski, who said: "At best the TTSS represents one possible step in the indirect Darwinian evolution of the bacterial flagellum. But that still wouldn't constitute a solution to the evolution of the bacterial flagellum. What's needed is a complete evolutionary path and not merely a possible oasis along the way. To claim otherwise is like saying we can travel by foot from Los Angeles to Tokyo because we've discovered the Hawaiian Islands. Evolutionary biology needs to do better than that."

The Aching Mystery: The Origin of Proteins

Throughout this chapter, we have almost forgotten who the quiet molecular stars of the show are—the unsung heroes in our task of tracking down how complex systems arose. Who are these stars? I'm referring to the individual pieces of cellular machines and systems, the proteins. When building a flagellar motor, the cell needs forty different proteins, each of them being different in its chemical code blueprint and in its amazing folded 3-D structure. Let me review the wonder of these remarkable workhorses. My goal is to explain why each and every one of them is an incredible *microuniverse of complexity*. Any single protein, on its own, is overwhelmingly implausible on any naturalistic theory. Each protein, in effect, strikes back at Darwin and vindicates Behe over against his critics.

Every protein molecule is a chain of smaller units, the chemical letters called amino acids. These small molecules, typically made up of about two dozen atoms (with some a bit less), come in twenty different kinds within proteins (many others have been synthesized, which are not used in proteins). Biology texts compare these twenty different amino acids with the twenty-six letters of the English alphabet. As I said, they are strung together in cells to form long protein chains, often several hundred units long, which quickly fold into a uniquely shaped final structure of a protein. Without proteins, we wouldn't have a functioning body. Proteins furnish what's needed for almost all of the structures and most of the chemical functions of all living things! They are the stuff of life.

At an ID seminar in Paris, I sought to model protein structure. I brought bags of decorative pompoms of various bright colors (the colors stood for the twenty amino acids). A tailor next door to our hotel sewed the pompoms together on a long thread. The pompom chain served as a crude model to show what proteins look like. I explained that real proteins are much longer than my pitiful model—they are incredibly long chemical words that, after assembly, begin to twist around themselves in the watery environment of the cell.[33] As they scrunch together, the chain eventually settles down into its precise final fold—a preprogrammed 3-D shape. You can't get a folded protein unless you have a specified sequence of individual amino acids. Many spots in the chain are very finicky. To illustrate these finicky positions, let's assume I have a protein with a hundred amino acids. Through testing, I find that only one amino acid works at certain spots: two, fifteen, thirty-six, seventy, and eighty-one on my protein chain. At other positions, there is more flexibility. For instance, at position fifty in the chain, we might find that five of the twenty will do okay.

At any rate, the twenty different proteins in the blood-clotting cascade, or the forty proteins in the bacterial flagellum, are each formed the same way, as a twisting chain of usually at least one hundred—and often up to a thousand or more—amino acids that are ordered in a careful sequence. Each protein is a specially sequenced chain that resembles a *gigantic chemical word*, since even short proteins (with very few exceptions) are at least one hundred units long.

One such protein that I'm depending on to type this chapter is *collagen*, the main structural protein by which an amazing number of tissues and structures in my body are built and supported. Our skin tissues are rich in collagen. This protein gives my fingertips solidity yet flexibility as I type this sentence on my computer's keyboard. Even the bones and teeth we depend on are built with the help of a collagen matrix. Collagen, in turn, depends upon a precise sequence of amino acids to enable its triple-helix bundle to be produced, with the help of vitamin C. Of course, this is just one protein among thousands of different proteins we depend on—including the special proteins called *enzymes*, which perform the trick of speeding up chemical reactions. Without enzymes, we would be dead.

As Michael Behe traced the twenty steps of the blood-clotting cascade, or the multiple steps of the vision cascade, every step depends upon a uniquely programmed and folded protein. Note that the shapes and precise surface contours of proteins *are crucial for function since they work together in a mutually regulated sequence*. In the vision cascade, they trigger those electrical signals (across millions of rod and cone cells in my retina each second) that result in me seeing the words of this sentence appear on my computer screen and allow you to see words on the page.

I could walk through dozens of other complex systems, but I think I've made my point. The foundational question in confronting irreducible complexity is this: *how did just one given folded protein arise* at the right time, with the right shape, from a random shuffling process, to be added to a growing system? It is certainly plausible (as biologists tell us in their protein-creation story) that an extra copy of a gene for one protein was duplicated, ready to be worked on by mutation, but that's just the start. *What is the likelihood that this unneeded gene copy had its letters shuffled around until it successfully produced another relevant folded protein? The answer: Darwinists don't have a clue. No biochemist has the answer.* The only estimates we have are horrendously remote, and so this is not an encouraging area for Darwinism, to say the least. Defenders of naturalistic evolution are hopeful that the mutational distance from one functioning protein to another distinctly folded protein is fairly short—say a few (two, three, or four) mutational steps will do the trick. Yet recent work by ID biochemist Doug Axe[34] *seems to show just the opposite*. Such folded proteins are not like little islands in an archipelago, allowing a short leap from one functional island to the next. Rather, proteins are so isolated from each other in their sequences that they have been compared with islands that are light-years apart from each other in the mathematical multidimensional "phase space," which models the space proteins would have to move through to be mutated and morphed successfully. As a result of these findings, Darwin's nineteenth-century theory of origins seems rendered obsolete through the twenty-first-century research into the staggering complexity of individual proteins.

It comes down to this: What confidence do we have that the cell will be producing new and structurally relevant proteins for a developing motor to co-opt and fold into its mechanical grasp? Is there a reliable supply? To be blunt, there is no empirical basis to expect any such supply. Nevertheless, Darwinists seem never to question such a steady supply of novel proteins. Such a ready supply is assumed; it is a *given*, a relaxed assumption. ID theorists balk not just at thirty new proteins arising (to carry a TTSS injector onward to a flagellum). ID finds very implausible *even one new relevant protein arising—any particularly useful and precisely shaped protein coming into existence by chance.* At the 1992 Darwinism Symposium, Behe tackled the improbability of deriving even one protein with proper shape and thus potential function. The probability is one in 10^{78}! The blithe response to Behe's paper came from a Princeton biologist, Leslie Johnson. She said, in effect, "No problem! I invent new foods every day in my kitchen—new fajita combinations, a new concoction called 'ginger beer,' and much more. Proteins are just new combinations of preexisting materials!" This reply not only lacks rigor, it seems astoundingly (and revealingly) irrelevant.[35] The flagellar motor requires not one but forty such highly implausible productions.

Kenneth Miller virtually skirts this issue; he writes as if new proteins are evolving right and left in the cell—a happy and bountiful waterfall of new precisely folded chains. It is as if the molecular world of a cell is a big, wonderful, protein-popping machine, spewing out a steady stream of new folded proteins—much as one sees the popcorn snapping and popping inside the Plexiglas cabinet in a movie theater. When facing the implausibility of evolving even a single novel protein for any irreducibly complex system, Miller's co-option argument is like a backpacker with tennis shoes struggling to climb a fifty-mile-high vertical cliff. Now add forty such cliffs, back to back—a two-thousand-mile-high vertical cliff. Any assumption of the cell as a *wonderful protein popper* is rooted in facile imagination, not rigorous science.

Thus we have descended downward to a more basic level, where we encounter the irreducible complexity of the combined letters that program the precise shape of a single folded protein. If nature-driven processes cannot form new and relevant proteins from the ceaseless activities of cellular life, then the evolution of an irreducibly complex system could never get started, let alone completed. The molecular heroes at this level have a powerful lesson to teach us all.

Jonathan Wells and *Icons of Evolution*

The Battle over Textbook Misinformation

The Proofs Go Sour, Very Sour

One of the main goals of this book is to explore and understand the deepening plausibility crisis that has enveloped Darwin's grand theory of "descent with modification"—a process driven entirely by the forces of nature. This crisis ultimately pivots on a few crucial questions, such as this one: how do we in the educated general public—the vast throngs of us nonscientists—come to know *for sure* that *nature, all on its own,* has produced the marvelous variety and complexity of living things spread out across our shimmering globe? Or to make it more succinct: how do we know that the Darwinian theory of macroevolution is true? When Jonathan Wells published his *Icons of Evolution* in 1999, he was setting out to explain and analyze ten of the top textbook proofs, including the peppered moth observations, which purportedly answered these key questions and provided convincing proof. These ten showcases of supporting evidence are called "icons" because they appear as pictures, diagrams, or other images in the lushly illustrated biology textbooks used in high school and college biology classes.

Icons of Evolution shows that each of the ten icons is highly problematic—their presentation in textbooks is plagued with misinformation. These sweet proofs have gone *sour*, and yet students are given no hint of this. What is served up, instead, is a distortion of the evidence for evolution. *Icons of Evolution* has sparked furious controversy across the U.S. and beyond. It has shocked readers from all walks of life. Many who read the book have felt scandalized and betrayed by science's failure in this basic area. Most importantly for U.S. education, Wells's book helped trigger a process of reevaluation that led to changes in state school board policy in several states.

Three years after the book was released, Coldwater Media published *Icons of Evolution*, an hour-long documentary that extended the reach of the book much wider. The video adds one major thread not found in Wells's book: the story of the persecution of Roger DeHart, a biology teacher in Washington State. Incredibly, under pressure from Eugenie Scott, the NCSE, and the ACLU, DeHart's superintendent ordered him to stop bringing in *supplementary* material from prestigious, peer-reviewed journals about problems with a few of the icons![1]

The *Icons of Evolution* video has worked well in tandem with Wells's book; the book's impact was probably quadrupled or more through the showing of the *Icons of Evolution* video. Together, they have made a deep and permanent mark on the origins debate. One major ID success story of concrete (if fitful) progress in the public arena is found right here, in the reverberations and reform triggered by Jonathan Wells.

Yet the unleashing of *Icons* was not without great cost. Wells endured some of the most vicious and blatantly false accusations of any ID scholar, some of which are still not retracted as of this date (early 2006), even though those responsible have been provided clear documentation of their distortions years ago. As Wells puts it,

> It seems that I am guilty of the one unforgivable sin in modern biology: I am openly critical of Darwinian evolution. In *Icons* I pointed out that the best-known "evidences" for Darwin's theory have been exaggerated, distorted or even faked. I argued that a theory that systematically distorts the evidence is not good empirical science—perhaps not even science at all. In fact, Darwinism has all the trappings of a secular religion. Its priests forgive a multitude of sins in their postulants—manipulating data, overstating results, presenting assumptions as though they were conclusions—but never the sin of disbelief.[2]

The attempts to tarnish Wells's reputation as a competent biologist and critic of neo-Darwinism are numerous, and some are egregious. One review of *Icons* in *The Quarterly Review of Biology* accused Wells of doing no experiments and publishing no peer-reviewed research during his postdoctoral work at Berkeley.[3] Wells's supervisor, Berkeley biochemist Carolyn Larabell, wrote a strong letter to the publisher of the review. Larabell unleashed a stinging repudiation of both claims. She pointed out:

> Both of these claims are false. Dr. Wells and I performed numerous experiments together in my laboratory at Berkeley while he was a post-doc. That research resulted in two peer-reviewed papers to which we contributed as coauthors. Some of our work has even appeared in a textbook on developmental biology. [References are footnoted in the letter.]
>
> I am surprised that *The Quarterly Review of Biology* would publish something with so little regard for truthfulness and professional decorum. The false claims of Padian and Gishlick unjustly damage not only the reputation of Dr. Wells, but also—indirectly—the reputations of those who worked with him. It seems to me that a retraction is in order.[4]

The magazine replied that its policy is not to publish retractions.

I would like to expose in greater detail other low tactics used against Wells, but I don't have room, and it would distract from the scientific issues, which are paramount. I refer the reader to Wells's must-read commentary "Critics Rave Over *Icons of Evolution*," available at www.Discovery.org.

In this chapter, I will (1) explain what Wells's overall project was and review briefly his ten icon critiques, and (2) select one icon for close scrutiny, to see how the controversy has played out since 1999. Finally, (3) I shall dissect a fascinating response to Wells from Eugenie Scott, which reveals the assumptive state of the Darwinian mind.

It will be profitable to list all ten questioned "proofs of evolution" that make up the chapters of *Icons of Evolution*. Below are the ten key points, as summarized by two of Wells's strongest opponents, Barbara Forrest and Paul Gross, in their caustic book *Creationism's Trojan Horse*. These ten points comprised—in tone at least—a refreshing window of emotional calm within their river of contemptuous rhetoric.[5]

> 1st Icon: Abiogenesis—life from nonlife—the synthesis of organic (biologically important) compounds, building blocks of life, from simple, nonliving chemicals, in the famous [Stanley] Miller-Urey experiment of 1953. That experiment doesn't work, says Wells, if it is done properly; therefore, it is wrong about the origin of life and thus no support to Darwinism. To teach about it is fraud.

Overall, this is fairly accurate. Yet the last sentence, "To teach about it is fraud," is a subtle distortion. On the contrary, Wells makes it clear that he wants teachers to maximize their teaching about it, if they will merely place it in its historical context of questions and problems brought up repeatedly by scientists since 1953. Also, nowhere in his chapter on Stanley Miller and Harold Urey does the word *fraud* appear, nor does he bring up this topic later when "possible fraud" is discussed.[6] Thus, Forrest and Gross's comment distorts Wells's remarks. In the appendix of *Icons of Evolution* where biology texts are graded, it is urged that textbooks, when discussing the Miller-Urey experiment, make it clear that the experiment "is probably irrelevant to the origin of life because it did not simulate conditions on the early earth." Wells will give an "A" grade if a textbook "mentions the controversy over oxygen in the primitive atmosphere, and includes extensive discussion of the other problems faced by origin-of-life research, acknowledging that they remain intractable."[7] Is this criterion unreasonable? In light of the complex problems of this field, one would assume such candor is very much in order. The Miller-Urey apparatus icon and the entire origin-of-life field—also called abiogenesis or chemical evolution—will be discussed further in chapters 8 and 9.

> 2nd Icon: The (Darwin's) "tree of life," that is, the evidence of a multiply branched descent with modification of contemporary and more recent species of animals from fewer, common ancestors in the past. Wells denies that there is any evidence for this, discounting in the process all of molecular phylogeny and insisting that there is no basis for assuming the existence of ancestors to fauna of the Cambrian "explosion."

Again, even though Forrest and Gross's coverage is generally fair, they subtly distort Wells at this point. Wells makes it clear that the field is *overflowing with evidence, but it is odd, conflicting, and has engendered rampant confusion*, leading to "tree diagrams" that much more resemble crisscrossing "thickets." He says: "Inconsistencies among trees based on different molecules, and the bizarre trees that result from some molecular analyses, have now plunged molecular phylogeny into a crisis."[8]

> 3rd Icon: "Homologies" of structure, for example, the close similarities of all the bones of vertebrate limbs, or [of] the structures and amino acid compositions of proteins with related functions from widely separated taxa. These similarities, commonly held to be evidence of descent from common ancestors, are not that at all, says Wells. His treatment of it makes the (universal) use of homology in the evolutionary sciences nothing more than arguing in a circle.

The authors have summarized well; this is commendably accurate and straightforward.

> 4th Icon: Haeckel's drawings of vertebrate embryos. These, Wells asserts, were faked. Moreover, the evolutionists know it and have known it for a

> long time. But until very recently they have said nothing about it *to protect these drawings as key "proof" of Darwinism*. Haeckel's drawings are still reproduced in textbooks.

This statement is a good summary of Wells's discussion; again the authors are to be commended for their accuracy. I will discuss this icon in more detail below.

> 5th Icon: Archaeopterix. This fossil form is commonly cited as proof of evolution, a "missing link" between the dinosaurs and modern birds. But it is not a missing link, Wells argues; it is not an ancestor of living birds, and it is therefore no proof of evolution.

Again, there is no problem here.

> 6th Icon: The celebrated case of the peppered moth. This famous example of Darwinian selection operating to powerful effect in nature, over short time-intervals, is a fake, Wells says, because, among other reasons, the photographs of moths resting on tree trunks, used in many textbook accounts, were staged (the moths were glued on). So the phenomenon for which these moths are famous (industrial melanism, protective coloration) is no support for the Darwinist idea of natural selection.

I could nitpick on their summary, but overall, the authors have summarized well. They are on a roll—four items in a row with no serious distortion!

> 7th Icon: Darwin's finches—another famous and often cited case of rapid natural selection in the wild. It isn't so, says Wells, for several reasons, most notably that observed population changes in beak morphology were not a result of macroevolution.

A reasonably good summary, but it's worded a bit strangely. It would be clearer simply to say: the changes in finch beaks *do not represent* a case of macroevolution; they are a case of microevolution—cyclical variation in tweaking existing structures—which has never been questioned by ID theorists.

> 8th Icon: Fruit flies with four wings. Says Wells: the case of a four-winged fly, appearing spontaneously in a species of two-winged insects (flies are "Diptera": two wings), is not, as claimed in some books, evidence for a neo-Darwinian mechanism of evolution.

Well done!

> 9th Icon: Fossil horses and directionality in evolution. Almost everything said and taught (illustrated in textbooks) on the evolutionary lineage of horses, a paradigm case, is wrong, says Wells, because evolutionists have rigid, materialist preconceptions about such phenomena as directionality and purpose.

It is a bit puzzling that this description leaves out the most important topic in this chapter besides horse evolution: the assumed *creative power of natural selection*. Perhaps the authors want to keep Wells's stiletto critique of the natural selection off the radar screen. Contrary to the impression here, Wells actually applauds the switch from orthogenesis description—straight-line evolution—of the horse lineage to a more branching description. It is the underlying scientific rationale for this switch that is brought into question.

> 10th Icon: Hominid evolution and humans. The gap between the anthropoid apes and humans is not really filled, says Wells, by the many hominid fossil sequences presented today as the human lineage, for, among other reasons, there was once a fraud in the business ("Piltdown man"). The experts long failed to recognize this; besides, there are constant disagreements today among them about what fossil species, among the scores now known, are ancestral to which others and about what the fossil sequences actually mean. Paleoanthropology, in other words, is untrustworthy on questions of human antiquity.

This description is reasonably accurate. The discussion of Piltdown, however, is not as simplistic as they present it—"there was once a fraud in the business"— nor was it very prominent. In a chapter with sixty paragraphs, just five of them discuss Piltdown.

Ernst Haeckel's Embryos in Focus

We now leave the survey of icons and zoom in on one of the most controversial: Ernst Haeckel's drawings of eight species—all vertebrates—passing through three stages of embryological development. This constitutes Wells's fourth icon. This icon is personal to me and my late father[9] because the famous charting of embryos played a key role in his cementing my own evolutionary belief during high school days. My father first showed them to me as a freshman studying biology in 1965. I'll never forget when he pointed out these impressive drawings, showing the similar structures—including tails and "gill pouches"—shared by human, chick, hog, tortoise, and fish embryos. They were printed in his Princeton textbook on organic evolution from the late 1920s.

By this time I was strongly convinced of both evolution and the power of its engine, natural selection. For me, the embryos were the evidential clincher and confirmatory slam dunk. I thought, "Wow—proof positive!" I asked my father to let me take his book into Doc Wynkoop, my biology teacher, who then was covering evolution. Doc was pleased that I brought in the additional evidence, and we showed the drawings to all the class. Little did I know that these drawings were distorted—and so seriously that Haeckel in his day was accused of fraud. In fact, "charges of fraud abounded in his lifetime."[10]

The *importance* of the embryological evidence, as crucial support for Darwin's general theory, is well captured by Wells in his response to the critics:

> Aware of the problems with the fossil record, Darwin thought that the best evidence for his theory came from embryology. He believed that early vertebrate embryos "are closely similar, but become, when fully developed, widely dissimilar." He concluded that this was not just evidence for common ancestry—it was "by far the strongest single class of facts in favor of" his theory. In the 1860s, German Darwinist Ernst Haeckel made drawings of vertebrate embryos to illustrate these "facts." Yet (as his contemporaries pointed out) Haeckel faked his drawings: Vertebrate embryos actually start out looking very different, then converge somewhat in appearance midway through development before becoming more different again as adults. Haeckel misrepresented the midpoint of development as the first stage, then he distorted the embryos at this point to make them look much more similar than they really are.[11]

If one turns to Wells's book itself, *Icons of Evolution*, the problems of these embryo drawings are not minor; they are multiple and major. For one thing, Haeckel only used five of the seven vertebrate classes in his drawings, and four of them (half of the eight total) are from one class—mammals! For a sample amphibian, he used a salamander rather than a frog, whose embryo looks much less similar to the other embryos in its development. In other words, he cherry-picked his evidence. Most importantly, it was as early as the 1890s, over one hundred years ago, that biologists knew "that Haeckel faked his drawings; vertebrate embryos never look as similar as he made them out to be. . . . Although you might never know it from reading biology textbooks, Darwin's 'strongest single class of facts' is a classic example of how evidence can be twisted to fit a theory."[12]

Whether Haeckel "was guilty of fraud . . . there is no doubt that his drawings misrepresent vertebrate embryos," says Wells, who later adds that when "Haeckel's embryos are viewed side-by-side with actual embryos, there can be no doubt that his drawings were deliberately distorted to fit his theory."[13]

This empirical failure is famously vouched for by a British embryologist named Michael Richardson, who published a new study in 1995 that reexamined Haeckel's drawings. Richardson concluded that these "famous images are inaccurate and give a misleading view of embryonic development." This was followed by a 1997 study by Richardson and an international team of experts that compared the embryo drawings with actual photographs of embryos from all seven classes of vertebrates. Again the results agreed that "Haeckel's drawings misrepresent the truth."[14] Wells reports: "In the March 2000 issue of *Natural History*, Stephen Jay Gould noted that Haeckel 'exaggerated the similarities by idealizations and omissions,' and concluded that his drawings are characterized by 'inaccuracies and outright falsification.'" Wells adds that when *Science* interviewed Richardson around the time his research was being published, he leveled with the writer: "It looks like it's turning out to be one of the most famous fakes in biology."[15]

To understand how important this case of fakery was in terms of a major argument for evolution, let's turn back the clock to the late 1800s. Here we find delicious irony in embryonic evidences. The father of modern embryology in the 1800s, Karl Ernst von Baer, remained opposed to Darwin's theory to the end of his life,[16] but Darwin continued to cite the evidence of embryology, published by von Baer, as support for his theory. In spite of the problems with Haeckel's embryos, by the early decades of the twentieth century when my father was hearing this evidence at Princeton, Haeckel's famous formulation had hardened into a dogma called the "biogenetic law," famous in its three-word poetic capsule: "Ontogeny recapitulates phylogeny." Translated, this means that the stages of development of an embryo (ontogeny) repeat and sum up the highlights (recapitulates) of the stages of that creature's evolutionary history (phylogeny). For example, the Haeckel embryos were thought to show "gill slits" or "gill pouches" in human or other mammalian embryos. (I recall my father enthusiastically pointing this out.) The tail of a human embryo looks similar, in the drawings, to the tail structure of other embryos. So, humans supposedly passed through key evolutionary stages during their development. Unfortunately for Darwin, Haeckel, and the biogenetic law, it turns out that the gill slits are neither gills nor slits. These structures are now called pharyngeal folds (or pouches), and in humans they never become lung tissue but rather glands, parts of the inner ear, and other structures.

By the 1960s the biogenetic law was relegated to the dustbin of biological history. I heard the renowned anthropologist Ashley Montagu admit this in a 1980 taped debate on the Princeton campus.[17] Unfortunately, the law lives on, zombielike, in a watered-down version today,[18] but Wells argues that the most devastating evidence of all, which practically crushes the theory, is that the earliest stages of embryonic development are completely omitted. And yet right here, at the very earliest stages—cleavage, gastrulation, and beyond—where Darwinian theory would suggest that structures would be most conserved and most similar, we find the greatest differences and structural disparities between embryos! In summary, the evidence of embryo development tosses a hundred-pound monkey wrench into the machinery of Darwinian macroevolution. It is definitely not convincing proof that deserves to be "presented to the class" as Doc Wynkoop and I did in 1965.

Darwinists Strike Back at Wells

This chapter easily could be three times longer if I traced the sizable literature that counterattacked Wells on each of the embryos, seeking to show him as being "ignorant, stupid, or wicked."[19] Since I'm limiting myself to the single icon of Haeckel's embryos, I will quote frequently from Wells's lively response to one of the principal reviewers who attacked his handling of the embryos. This reviewer, Jerry Coyne, is a leading critic in print of Intelligent Design, and he wrote his review in *Nature*, the most prestigious science journal in the world. Coyne spent considerable space attacking Wells (attributing evil motives and connecting him with creationism generally) rather than dealing with the content of his book. Wells points out that Coyne's attack on his handling of embryology is "a daring move on his part, since I'm a vertebrate embryologist and he's a fruit fly geneticist."[20]

Coyne begins by repeating the standard textbook use of embryos as evidence for evolution: "As Darwin first realized, some aspects of vertebrate development—especially transitory features—provide strong evidence for common ancestry and evolution. Embryos of different vertebrates tend to resemble one another in early stages, but diverge as development proceeds, with more closely related species diverging less widely. This conclusion has been supported by 150 years of research." Then Coyne goes on to rebuke Wells's effort "to refute this mountain of work."[21]

Wells's response is pointed and devastating.[22] It deserves to be presented in full:

> Naturally, I would be grateful to Coyne for correcting me about this—if he were right. But his claim that vertebrate embryos are most similar in their early stages is dead wrong. As British zoologist Adam Sedgwick wrote in 1894, the claim is "not in accordance with the facts of development." Comparing a dogfish with a chicken, Sedgwick wrote: "There is no stage of development in which the unaided eye would fail to distinguish between

them with ease." It is "not necessary to emphasize further these embryonic differences," Sedgwick continued, because "every embryologist knows that they exist and could bring forward innumerable instances of them. I need only say with regard to them that *a species is distinct and distinguishable from its allies from the very earliest stages all through the development*" (emphasis in Sedgwick's original).[23]

Many other vertebrate embryologists have noted the same thing. In 1976, Dartmouth College embryologist William Ballard wrote that it is "only by semantic tricks and subjective selection of evidence," by "bending the facts of nature," that one can argue that the earliest stages of vertebrate embryos "are more similar than their adults." And in 1987, Canadian embryologist Richard Elinson wrote that early developmental patterns in frogs, chicks and mice are "radically different."[24]

So the "mountain of work" Coyne invokes actually buries his claim. But that doesn't seem to bother him, because . . . he acknowledges that vertebrate embryos are not most similar in their early stages: "Wells also notes that the earliest vertebrate embryos (mere balls of cells) are often less similar to one another than they are at subsequent stages when they possess more complex features." Like other evolutionary biologists, Coyne argues that the dissimilarity of early vertebrate embryos can be explained in the light of Darwin's theory, since "the earliest stages of vertebrate embryos show adaptation" to the conditions of their existence. Coyne even regards this as evidence for the theory: "Wells repeatedly fails to grasp the evidential value of phenomena [i.e., dissimilarities in early embryos] that can be understood only as the result of a historical process."[25]

The pathos and irony of the pattern of Darwinian doublespeak, the win-win situation whereby both early dissimilarities and later selective similarities work to the advantage of the reigning theory, is evoked with appropriate sarcasm:

So let me get this straight. Some of the strongest evidence for Darwin's theory is that vertebrate embryos are most similar in their early stages—except that they're not. But if we just interpret the embryos' dissimilarities in the light of Darwin's theory, they then have "evidential value."

Oh, now I get it! Darwin's theory wins no matter what the evidence shows. Apparently I was just ignorant of how evolutionary biology works.[26]

The other published Darwinian attacks on the remaining nine icons of Wells have much of the same flavor. Either Wells doesn't have his facts straight (and in each case, he quietly corrects the attackers on their misconstrual of the facts), or his basic logic of evaluation is skewed. I invite the reader to peruse the "Critics Rave" reply of Wells on the Discovery Institute website. It represents the single most thorough and in-depth engagement in print by any ID theorist responding to his published critics. I just wish that another round were possible—the critics reply to "Critics Rave," and Wells responds to their reply. But it's time to move to the spot where Eugenie Scott explains why the entire approach of Wells is simply wrongheaded.

Eugenie Scott Strikes Back

Eugenie Scott is the director of the National Center for Science Education, so her job resembles that of Max Mayfield, director of the National Hurricane Center. Scott tracks all storms (or early indicators of potential storms) of anti-Darwinian controversies in the schools, universities, and governmental institutions of the U.S. Her goal is not only to warn the pro-evolution constituency of these threats but to seek to block and thwart them at every turn. So Scott attempts to do what Mayfield cannot. She coordinates the halting and smothering of the storms of skepticism.

When *Icons of Evolution* appeared, Scott gave a lecture on the ID threat at the University of California at San Diego. Casey Luskin, a founder of the IDEA Club on that campus, attended Scott's talk and heard her describe *Icons of Evolution* as a "royal pain in the fanny" for Darwinists. I quoted that comment in chapter 1 of *Doubts about Darwin*, and it is appropriate to mention it again as background. I begin with the opening thrust of Scott, which compares evolution to other branches of modern science such as atomic physics. Scott begins:

> If someone were to charge that textbooks present atomic theory using evidence that is erroneous, misleading, and even fraudulent, and that we should therefore question whether matter is composed of atoms, eyebrows would be raised—at least at the accuser. If someone further claimed that distinguished physicists crassly participate in this fraud to keep the research dollars rolling in or to promote a materialist philosophical agenda, scientists would be angry at the attempt to besmirch the reputations of respected scholars. And if the same person proposed that citizens should encourage local school boards to insert anti-atomic theory disclaimers in science textbooks, discourage Congress from funding research in atomic theory, and lobby state legislatures to restrict its teaching, it is doubtful that such exhortations would receive much attention.
>
> Such would be the fate of Jonathan Wells's call to arms in *Icons of Evolution*, if biological evolution were not substituted for atomic theory in the above scenario. . . . Unlike atomic theory, evolution has obvious theological implications, and thus it has been the target of concerted opposition, even though the inference of common ancestry of living things is as basic to biology as atoms are to physics.[27]

Through this introduction to her review, Scott seems to suggest that the reader view Wells's book as something bordering on lunacy. The argument works only to the extent that the reader assumes that Wells's critique of the icons is at least somewhat equivalent or conceptually parallel to the insane critique of atomic theory. She makes clear that in her mind the only real difference is that atomic theory is nontheological in its implications, while Darwinian evolution is notably theological in implication. Clearly, for the analogy to work in Scott's mind—and in the mind of the readers—one must view the inherent evidential foundations as equally solid in the two theories. Likewise, one must take it as a given that the explanatory power is equally cogent when comparing the two. Only "theological implications" are at issue; these and these alone separate the two.

Yet of course, that bare supposition of Scott's (and her readers) is the exact point at issue; it is the point at which *Icons of Evolution*'s critique is delivered. Wells is not saying merely that there are problems in *how the icons are presented*. Rather, he asks whether the icons, as properly understood in light of the evidence, give any factual support at all for the creative power of natural selection itself, or even for the tree of life (the doctrine of common ancestry).

At one point in her review, Scott implies that the scholarship in *Icons of Evolution* may be technically correct, but the overall effect of the book is to simply mislead the reader: "Wells presents a systematically misleading view of evolution. Individual sentences in *Icons of Evolution* are usually technically correct, but they are artfully strung together to take the reader off the path of real evolutionary biology and into a thicket of misunderstanding."[28] It seems that the worst thicket that Wells leads the reader into is the one labeled "Cambrian explosion," which is briefly discussed in chapter 3—on the icon called "Darwin's Tree of Life." Scott complains,

> The Cambrian explosion is supposed to be a "serious challenge to Darwinian evolution" because "phyla and classes appear right at the start." Wells is wrong to claim that the Cambrian appearance of major body plans supposedly puts paleontologists into a tizzy; actually, they regard it simply as a phenomenon yet to be explained. Unexplained is not unexplainable. More misleading to nonscientists is the implication that most modern phyla and classes occur in the Cambrian, which doesn't hold true for either animals or plants. Wells neglects to mention that insects, amphibians, reptiles, birds, and mammals are all post-Cambrian (and even Cambrian "fish" are problematic).[29]

Wells responded to Scott's jab on the Cambrian: "I never implied that the Cambrian explosion included plants; indeed, there is no such thing as a plant phylum (the major groups of plants are called 'divisions'). Nor did I ever imply that 'insects, amphibians, reptiles, birds and mammals' appeared in the Cambrian explosion—though the phyla to which these organisms belong (arthropods and chordates) did appear abruptly in the Cambrian. So what Scott criticizes is something I never claimed. She can't fault me for stuff I did write, so she tries to fault me for stuff I didn't write."[30]

It is appropriate that this chapter is ending on fossil evidence for macroevolution—the topic of our next chapter. Many evolutionists say that the most direct proof of the tree of life is the pattern of bones, shells, teeth, and other traces of ancient life embedded or imprinted in the rocks. Is this evidence a slam dunk for macroevolution? Let's have a look.

7

Fossils and the Battle over the Cambrian

The Iceberg and Its Shining Tip

Does the fossil evidence as a whole support the Darwinian story of macroevolution? How do Darwinists handle the infamous gaps in the fossil record, where transitional forms are notoriously scarce (or nonexistent)?[1] Most crucial of all: *What exactly are the most important gaps? Have they been steadily filled in over time, or have they persisted through the decades since Darwin?* Evolutionists and ID theorists, with few exceptions, provide radically conflicting answers to these questions. Moreover, this cluster of scientific questions is at the heart of the battle over design theory and Darwinism. Even though this chapter cannot begin to exhaust this vast and exciting topic, we at least can shed light on the structure of the contemporary tussle over ancient rocks, which pivots on whether fossils loom as a huge plus—or an embarrassing minus—for Darwinian theory.

As we flew over the main battle scenes where the exposés of Jonathan Wells are reshaping the current debate, I deferred the question of fossils to its own chapter. Perhaps this fossil focus can be partly chalked up to my own enduring fascination. In the past few years, my wife and I have had the pleasure of touring two of the world's finest fossil collections: the American Museum of Natural History in New York City and London's unforgettable British Museum.

We spent an entire day in each and, by prearrangement, my wife went on her own high-speed survey of all major galleries while I dawdled and gaped at the astonishing dinosaur fossil exhibits. Maybe it's the kid in me, but I never tire of seeing the bones, teeth, shells, and other hard parts of long-extinct animals that once lumbered (or skittered) across ancient swamps and plains.

Quite often, when I'm invited to a university to give a talk on the scientific controversy over Darwinism, I display my genuine amber drop with an embedded insect (purchased when we lived in the Dominican Republic), a gorgeous skull from an extinct type of mammal called an oreodon (an extinct camel-like mammal), and a spectacular trilobite from Morocco with lovely loopy antennae. Usually I begin with PowerPoint slides that survey the fossil evidence. My reason is simple: fossils open a unique window on the past that gives them extra high stakes as a test of Darwin's theory. Above all, I emphasize the colorful debate over the Cambrian explosion—the sudden burst of an incredible variety of complex animals in the fossil record about 530 million years ago. The name "explosion" is used widely in the literature of professional paleontology in describing this dramatic fossil debut (although one leading expert, Simon Conway-Morris, prefers to put quotation marks around "explosion").[2] This striking and puzzling phenomenon deep in Earth's ancient rocks is a saga so crucial to this debate that it must be retold here, complete with recent twists and turns.

A Quick Sketch: Problematic Iceberg and Treasure Trove

Before the Cambrian storytelling begins, let's start with succinct sketches of the drastically contrasting positions on the issue of fossils in general in relation to the Cambrian. Next, I'll expand the points of view on both sides, showing how their key arguments are erected. Last, I'll take the reader on a brief "Magical History Tour" of the ongoing controversy swirling around the Cambrian discoveries.

In a nutshell, in the view of design theorists, the Cambrian explosion stands as the *tip of the iceberg* of Darwinism's fossil problems. In other words, the problem of extreme scarcity (or total absence) of transitions is seen as the iceberg, and the Cambrian is merely the most extreme and astonishing display of that overall pattern. Besides being the tip of the iceberg of Darwinian headaches, the Cambrian fossils are viewed as constituting the crucial answer to the two questions that I placed in italics in the opening paragraph: (1) *What are the most important gaps?* Answer: the ones found in the Cambrian explosion, where we find not just gaps between slightly different forms but fossil chasms between different phyla that abruptly appear in the rocks. (2) *Are these gaps being filled in, or are they persisting?* Answer: the Cambrian gaps are persisting, with a defiance and stubbornness that is now legendary. What's worse, those chasms are not just enduring; they are steadily increasing in number through discoveries of new bizarre creatures made in recent decades. For ID's strategic purposes, which call on concentrating one's critical fire on the most crucial and vulnerable point on the opposing side, most research and writing has focused on that gleaming Cambrian tip of the iceberg and not the rest of the fossil gaps.

On the other hand, most evolutionists view paleontology and its rich fruit—countless millions of rock specimens from around the world, recording over 200,000 species of past life—as a bonanza, a tremendous *treasure trove* of support for their overall picture of how life developed. Above all, Darwinists stress that this trove of fossils has traced an undeniable *progression of species*—for example: bacteria appear before complex animals. Once multicellular life appears in the Precambrian and becomes well rooted in the Cambrian, there are still no bony fish at that stage—such as bass, perch, haddock, or trout. In the Cambrian, we find just a few very primitive (almost unrecognizable) chordates, which at best seem to be proto-fish. Then later, at higher geological strata, the familiar bony fish appear, followed in succession by amphibians, then reptiles, then mammals and birds, and finally apes, which lead onward to man. Incidentally, the majority of ID theorists accept the conventional dating of the geological ages, so this sense of *fossil succession* is not a disputed point for them. What *is disputed* by most scientists on the ID side is whether a given new form arose from earlier animals—or was ancestral to later groups—*and whether natural processes drove the changes.*

From the Darwinian perspective, the Cambrian mystery is (at most) admitted as a nettlesome problem to be worked on—which eventually will be solved. It is not a defeater for Darwinism. Above all, the Darwinian strategy is that the Cambrian mystery should not be allowed to dominate the conversation. Accordingly, it is almost never mentioned in presentations of fossil support for evolution. (This is commonsense rhetoric: why mention a glaring mystery when persuading the public? Above all, stay on the offense. Don't go on the defense unless necessary!)

Elaborating the Fossil Standoff: ID Speaks Out

As I proceed to elaborate on the two sides, I'll start again with the side of Intelligent Design. The best way to capture ID's perspective on fossils is to return to basics and reemphasize the two large patterns now visible in the record of the past, which are found embedded in sedimentary rocks. These universally recognized patterns of fossil data were made famous by Stephen Jay Gould and his colleagues in developing a theory called "punctuated equilibrium." These are: (1) *sudden appearance* of new forms, bursting onto the scene without identifiable ancestors, followed by (2) *stasis*—a *persistence of form* or *structural stability* (a sort of "resistance to evolution").

For those unfamiliar with Gould's theory, punctuated equilibrium holds that most living things did not steadily and continuously evolve into new forms, as if one were moving up a ramp, but they evolved in more of a stair-step fashion. The stability of form (stasis) is pictured in the flat tread of a step, and the abrupt rise of a new species—the vertical part of the step—occurs very rapidly. Because of the rapidity of the quasi-jump to a new form, and because the change takes place in such a small, isolated inbreeding group, it would necessarily be a very rare event for one of the transitional forms to be fossilized. This was Gould's way of accounting for the extreme rarity of transitional intermediates. (In describing the rise of a new species as "rapid," Gould referred to geological rapidity: perhaps thousands of years—a mere eyeblink compared to the millions of years of stable, unchanging form, during the stasis periods.)

Design theorists make several key points from the pattern of *sudden appearance followed by stasis*: First, this type of pattern seems to be quite at odds with the picture we have from standard neo-Darwinism, which preserves the step-by-tiny-step scenario that Darwin expressed so well, leading to a living universe of life-forms undergoing inexorable and continuous change: "It may be said that nature is daily and hourly scrutinizing, throughout the world, every variation, even the slightest, rejecting that which is bad, preserving and adding up all that is good, silently and insensibly working, whenever and wherever the opportunity offers, at the improvement of each organic being."[3] This picture of continuous change seems profoundly at variance with the stair-step pattern of punctuated equilibrium. (In fact, Darwin made it clear that in his theory, nature can never make a sudden leap to a new form, an event sometimes referred to as a "saltation." In his writings, Gould discusses this huge hang-up of Darwin and seemed to be engaged in a lifelong struggle to "reform Darwin" and stretch the original theory so that it can, indeed, accept some role for modest saltation-jumps along the way.)

Second, Darwin understood the objection to his theory from the lack of transitions both in the living world and in the fossil record, and he clung heavily to one notion to explain this. His key for explaining this state of affairs was simple: the "steady extermination of intermediates" by newer and improved life-forms. This is sometimes called "competitive replacement" in modern discussions. Yet, according to paleontologist David Raup, the vast majority of extinctions take place in massive deaths triggered by global catastrophes and are *definitely not brought about steadily by competitive replacement*. Also, this replacement mode implies a typical scenario where, for example, a primitive proto-bat, symbolized as "A," is outcompeted and replaced by evolved proto-bat "B," which in turn is replaced by "C," and so on until we reach "Z"—the true modern bat. Yet again, that's not what the fossil record shows. The facts are well-known that the earliest bats appear to be very much "Z" level bats (fully modern looking) and that we can't find among the fossils any significant progression either up to that point or from that point onward (only variation occurs thereafter). Again, the competitive replacement scenario seems not to fit well with the ubiquity of sudden appearance and stasis.[4]

Third, and most important: even in Gould's scenario of evolution, one would expect that intermediates would still exist (and occasionally be found) to some extent, at least when crossing the bigger morphological chasms. For example: a bacterium doesn't go over into the corner of a large population of bacteria and then, after a few thousand years of lucky macromutations, evolve the body plan and genetic library to build a complex animal—such as a trilobite or a jellyfish. Such a drastic morphological jump is plainly preposterous. So, transitional intermediates would still be expected here and there in the fossil record when the gaps yawn wide, and the greater the distance one travels in changing the morphology (body plan) from one kind of animal to the other, the more intermediates one would expect.

Here is exactly where the Darwinian explanation seems to collapse. On the video documentary *Icons of Evolution*, which parallels his book of the same title, Jonathan Wells explains this fairly simply with an illustration, "If you think of Darwin's branching tree, with a common ancestor down here [at the bottom of the tree] and the different modern forms of animals up here [at the branch tips], you would have one form to begin with and then it would gradually diverge into slightly different forms, and more and more different, until you get the major differences that we see now. The problem with the Cambrian explosion is that *all these major differences appear together at the same time with no fossil evidence that they descended from this common ancestor*."[5]

Where evolutionists see animals of the greatest conceivable structural difference appear suddenly, such as the huge difference between trilobites (of the arthropods) and starfish (an echinoderm), they would have to assume that these two forms have descended from some other, simpler ancestral phylum (or phyla) in deeper rocks. Yet the pickings are slim for ancestors. The candidates, in underlying rocks, are merely single-celled life or sponges (in China, sponge embryos are now found in rocks immediately below the

Cambrian), or the featherlike Ediacaran animals (which are generally seen as unrelated dead ends, not ancestors). The popular evolutionist escape route, which says that the ancestors were soft-bodied and could not be fossilized, has been firmly blocked by the new findings of very soft sponge embryos beautifully fossilized in the Chinese Precambrian sediments. Of course, the normal expectation is that to bridge these gigantic structural chasms, we should find *more* transitional intermediates than in normal gap situations. *And yet, there are none at all.*

To summarize: the biggest changes in body structure would involve the most intermediates, and thus the bigger the chasm between two body plans, the more stepping-stones we should find among the fossils. But precisely the opposite is seen in the Cambrian. Just where we need more transitions to buttress Darwin's theory (between phyla), we find none. Are these gaps persisting through time? ID theorists reply, "*Absolutely*." And that's why they emphasize the Cambrian mystery that keeps getting worse over time.

Darwinian Cheerleaders

Nevertheless, in biology textbooks the macro-theory is presented as if its fossil evidence is in great shape. In these books, both at the high school and college level, authors display prominently the tree of life (the most common icon of Darwin's descent with modification, with all organisms coming from an original primitive single-cell life-form). They do not describe the tree of life as a theory but as a fact. Jonathan Wells points out that this "fact" of a single tree with common trunk is indeed controversial, even outside Intelligent Design circles.[6] And, unsurprisingly, the Cambrian explosion is rarely mentioned, except in a passing reference that gives no hint of the persistent mystery.[7]

As an example of the assertive tone of Darwinian discussions of fossils today, let me spotlight one champion of macroevolution, writing in the November 2005 special anti-ID issue of *Natural History*. The author, geologist Donald Prothero of Occidental College, was positive, even ebullient, about fossil evidence. He conveyed his bullish message in the title: "The Fossils Say Yes." Admitting the weakness of the fossil support that Darwin struggled with in his day, Prothero exults that "evidence supporting evolution has continued to mount, particularly in the past few decades."[8] His five-page article is rhetorically very clever. It's about as effective as you can get in making the best of a bad fossil situation. It is splashed with artistic re-creations of extinct animals that are held up as transitional forms along with a photo gallery of skulls linking primitive hominids with modern man.

In the text, Prothero argues that key fossil gaps are being filled in, and he describes several new findings since 1983 of transitional species that shore up the scenario of whales and dolphins evolving from ancient land mammals. He also touts the alleged transitional fossils between dinosaurs and birds, between reptiles and mammals, and between primitive hominids and our own species, *Homo sapiens*. As I expected, among the gaps being filled in there is absolutely no mention at all of the Cambrian chasms.

Thus, as we saw earlier, each side of the controversy has typically gone on the offense, pointing out the fossil data that vindicate its point of view. It is almost as if the two sides are not directly interacting on the same issues, but there is one topic where both sides do comment (with the Darwinists doing so in a defensive mode): the Cambrian explosion. This is the one topic too weird and massive to avoid. Now it is time to dig deeper into this amazing story.

A Magical History Tour of the Cambrian

No one has done more to open up the mysteries of the Cambrian explosion to public view (and to try to quickly solve it with his theory of punctuated equilibrium) than Stephen Jay Gould. To help us sense the wonder of the most massive of all the sudden appearances in Earth history, and to catch the sheer excitement of this event, I shall let Gould himself describe this phenomenon. I highly recommend his book on the topic, *Wonderful Life*, a tale of the Canadian discoveries that shocked the world, along with Conway-Morris's 1998 book, *The Crucible of Creation*.[9] But here I quote Gould in an article in *Natural History*:

> Cambrian seas teemed with life preserved in an abundant fossil record. But when geologists . . . studied rocks of earlier, Precambrian times, they found nothing organic—not a trace of anything potentially ancestral to the diverse assemblage of trilobites, mollusks, brachiopods, and other creatures in Cambrian strata. This geologically abrupt transition from blankness to a rich fauna including representatives of almost every modern phylum has been called, in well-chosen metaphor, the "Cambrian explosion."[10]

Above, I emphasized that this abrupt debut of the richness of the Cambrian fauna had no discernable connection with previous life. This raises a question: was there any life in the rocks below the Cambrian strata? For over a hundred years (from Darwin's day until shortly before 1960), paleontologists thought that the Precambrian rocks were completely devoid of fossils—like a blank sheet of paper. However, since the 1960s especially, scientists have learned that Precambrian rock does contain quite a few scattered traces of life—mostly single-celled microfossils, such as bacteria and blue-green algae. Also a rare and unrelated set of creatures called the Ediacaran fauna (in the Vendian era) have been noted and cataloged. Most appeared within 30 million years of the Cambrian era, and many of them look vaguely like feathers rooted to the sea floor. One oddball, the Tribrachidium, bears a resemblance to a Frisbee with an odd decoration reminiscent of a swastika (only with three arms instead of four). Yet, since the Ediacaran animals are not considered ancestral to those of the Cambrian, and since bacteria clearly do not evolve abruptly into complex creatures like starfish, the Precambrian discoveries have only heightened the mystery of the Cambrian explosion of life.

This remarkable fossil explosion down at the very basement of the fossil record has long been known and was even discussed in Darwin's day. In fact, this sudden appearance of the major types of life-forms in the lowest strata of rock, without any trace of origin from earlier progenitors, baffled Darwin greatly. To his credit, he even admitted in *The Origin of the Species* (1859) that this "is the most obvious and gravest objection which can be urged against my views."[11]

The Burgess Shale: Discovery and Rediscovery

Now, fast-forward to the eve of World War I. A celebrated American paleontologist, Charles Walcott, stumbled upon a fossil gold mine high in the Canadian Rockies in 1909. For seven years, digging summer after summer in a Cambrian formation called the Burgess Shale in British Columbia, he unearthed some of the most beautifully preserved specimens ever seen from that era. These priceless fossils soon came into the possession of the Smithsonian Museum in Washington, DC. This precious cargo was stashed in safe storage and then largely forgotten for fifty years.

In the 1970s, a trio of British scientists[12] based at Cambridge University began to study the Burgess Shale fossils, which had never been closely analyzed. They blew away the dust and began to scrutinize each unusual creature in the collection. Soon, they were puzzling over unexpected body features and were astounded to find that several dozen of these animals were major discoveries—true oddballs hitherto unknown in the Cambrian explosion. In fact, many Burgess fossils fit no known category of living things. A striking example is the ugly little monster called *Opabinia*. It had a streamlined body, a piper-cub tail, a head with five protruding eyes, and sticking out in front, a proboscis structure resembling a fire hose, with a grasping tip at its end. Such a creature was previously unknown among the already rich Cambrian explosion.

I could list and describe many other new Cambrian creatures that emerged from Walcott's Canadian gold mine, such as three personal favorites—*Wiwaxia*, *Marella*, and *Naraoia*—but here I must sum up. With the careful study and reclassification of the Burgess Shale oddballs in the 1970s and 1980s, the Cambrian explosion suddenly got significantly bigger. A *Time* magazine cover story on the Cambrian discoveries published in December 1995 gave an apt title: "Biology's Big Bang." Because many of these creatures seemed to fit no known phylum or class, the extreme disparity—the sheer difference in body plan from other Cambrian animals—is what struck paleontologists. Gould pointed out in *Wonderful Life* that the popular textbook idea of evolution producing a "cone of increasing diversity" in basic life-forms over the ages was suddenly turned upside down. That is, the base of the cone (representing the time of greatest diversity) is at the bottom of the fossil record, in the Cambrian era, not in the present. After that, among animals no new basic body plans emerge; they just sometimes go extinct.[13] The Cambridge fossil experts had shocked the world of science, but as their work wound down in the 1980s, they could not have dreamed that an even greater series of Cambrian surprises would explode—across the world in southern China.

Chinese Monsters Explode on the Fossil Stage

Dr. Jun-Yuan Chen is a tall, thin paleontologist who teaches at China's prestigious Nanking Institute of Geology. In the mid-1980s, one of his former students showed his professor a beautifully preserved Cambrian fossil he had dug up in southern China, near the Vietnam border. The fossil caught his attention, and Dr. Chen asked the student to show him where he had found it. The student led Dr. Chen to the spot near the village of Chengjiang. This village is one of many that cluster in a rolling agricultural countryside in Yunnan province, peppered with hills and lakes.

Chen and his colleagues immediately began digging in this area and were stunned with the rich finds they were unearthing. By the early 1990s, thousands of precious Cambrian specimens had been pulled from the fine yellow shale, which had preserved even the delicate soft-bodied Cambrian animals in exquisite detail. The student had led Dr. Chen into the greatest Cambrian fossil bonanza of all time—eclipsing even the amazing Burgess Shale fossils.

In the past twenty years as the digs at Chengjiang have spread to more and more sites, the Cambrian fossils unearthed so far have begun (again) to shake up the world of paleontology. In China, fresh discoveries—some of the weirdest new species yet—have poured out of the Chengjiang fossil beds as well. The most bizarre and fearsome of all the new Cambrian species was *Anomalocaris*. This nightmarish beast had a large main body shaped somewhat like a flying saucer, with a large, tapered tail lined with swimming lobes. Protruding from the top of its head near the front was a pair of large eyes on stalks, and reaching out in front was a pair of grabbing and feeding arms that vaguely resemble jumbo shrimp.

When Dr. Chen came to Florida in 1999, it was my privilege to travel with him for a week as he visited major colleges and universities and to see firsthand the spectacular Chengjiang fossils he brought with him. I recall especially the tiny yellowish fossil, about four inches in length, in which one could see in dramatic detail the entire body of an *Anomalocaris*. I asked if this was a full-grown specimen. Dr. Chen laughed, "Oh no, that is a baby. We have found some that are six feet in length!" Apparently this fearsome animal was the king of the Cambrian seas, since almost all of the other creatures found in the various Cambrian fossil beds around the world are less than six inches in length. I found myself secretly happy this sea monster had become extinct.

Summing Up: So What?

The Canadian and Chinese fossil discoveries together constitute one of the greatest modern stories of paleontology, and that story is still unfolding as you read these words. Each year more and more unique, never-before-seen fossil specimens are being dug from beds near Chengjiang. In every case, the new Cambrian animal forms *appear suddenly*, with no hint of transition from anything else, either contemporary or earlier in the rocks, and after their appearance they *stay the same*—they manifest *stasis*—until they become extinct and disappear from the fossil record in higher layers. So the same pattern I reviewed earlier is seen here again, except it is magnified and lifted to a loftier scale in the Cambrian. There are so many different phyla and classes appearing at once, it resembles a gigantic machine-gun burst. This deepening mystery is the most glaring section of Darwin's iceberg of fossil mysteries.

Therefore, the truth claims of Darwinism, say design theorists, face grueling questions from the Cambrian data; this constitutes a powerful test case. It is here where the Chinese fossil scientists have made some interesting contributions to ID's case. For example, Dr. Chen says on the *Icons of Evolution* video: "Darwinism is maybe only telling a part of the story for evolution." While visiting Florida universities, I heard Dr. Chen say repeatedly, near the end of his lectures, "Darwinism cannot explain the Cambrian explosion anymore. We need a new theory."

The Darwinists Strike Back

The Cambrian animals are very special indeed, and their sudden appearance remains deeply unsettling for any Darwinian scenario of nature-driven development. Darwinists are of two minds in replying to this scientific nightmare. One is to offer possible explanations or counterevidences, which muffle the explosion, or which suggest a "long fuse" period of development before the Cambrian, as opposed to the apparent "short fuse" impression one receives from the fossil evidence.

A second way of replying to the Cambrian mysteries is to ignore the problem and accentuate the positive—gaps where transitional forms are seemingly beginning to fall into place quite nicely. Earlier, I mentioned the enthusiastic article from Donald Prothero, "The Fossils Say Yes"—a short but clever piece that paints a scenario in which the fossil evidence is now pictured as turning in favor of macroevolution. But Prothero is not the only writer who portrays fossils as a plus for Darwin; ID's dogged opponent Kenneth Miller forcefully seconds the motion at every opportunity. You'll recall Miller as the biologist who perfected the verbal sledgehammer technique. He links his arguments with words that smash and bash the arguments of his ID opponents. His 1999 book, *Finding Darwin's God* (profiled in chapter 3), cleverly weaves verbal blows in with his chosen bits of fossil evidence. The result is a barrage of rhetorical savagery that is calculated to comfort Darwinist believers and discourage doubters. For example, replying to Phillip Johnson's revelations of weak fossil evidence of vertebrates, Miller zeroes in on a crucial oddball fish suggested by evolutionists as a good intermediate: the rhipidistian fish. He takes on Johnson's argument, which basically said: In spite of its weird lobe-fins, this fish actually gives no indication how it might eventually become a four-legged amphibian crawling from the sea onto the land. Johnson says that, lacking such evidence, we can dismiss its potential ancestral role leading to four-footed land animals, the tetrapods.

Kenneth Miller then tries to sweep Johnson off the stage, claiming that Johnson's arguments had been defeated by new evidence. He pointed to one end of the fish-to-amphibian ancestral branch, where an amazingly fishlike amphibian (*Acanthostega gunnari*) had been found to fit beautifully. Then, moving to the fish end of this same branch, he highlights a fish possessing a "fin with fingers, eight in number, just like the digits of the earliest tetrapods." These examples seem truly interesting (if far from convincing), and in a more scientifically cautious rhetoric, they would merely serve as food for thought—possible collateral support for Darwinian theory. But in the hands of Miller, these examples are transformed into deadly darts against ID. To get the feel of Miller's blended attack (minor bits of evidence with generous doses of sledgehammer rhetoric), let me quote his key sentences and italicize some of the hammer blows:

> When that is done [namely: seeing if objections are grounded in fact] Johnson's objections *collapse*. His claim of a missing mechanism is *easily refuted*, his hopeful misinterpretations of punctuated equilibrium *fall apart* under close scrutiny, and his assertion that the fossil record does not support evolution *is in error*. . . . Two important fossil finds, one on each side of the fish-to-amphibian transition, *have crushed his argument. . . . Every objection of Johnson's has been answered. The fossils have been found in exactly the right place, at exactly the right time, with exactly the right characteristics to document evolution*.[14]

Miller's argumentation is technically impressive. He has perfected a powerful, emotionally forceful type of *logos* that is useful for making the most of one's limited evidence (like a defense lawyer). Yet I continually get the impression, when I encounter such stretches of sledgehammer rhetoric, that the scholarly credibility of his argument is being sacrificed. Where are the careful explanations of the all-important natural mechanisms that plausibly accomplished these amazing transitions? He has none.

Where are the indications, either from studies in the world of nature or in the laboratory, that *animals possess this degree of plasticity of form*? Kenneth Miller gives no evidence whatsoever. As a rule, he seems to prefer to assume, throughout his book, that nature is reliably writing new genetic information, ready to be used in serving the needs of complex systems and new body plans, which are being reshaped. This glib assumption that new DNA happens in living things—just naturally, through natural selection—is in my view (and that of the ID Movement) the ultimate Achilles' heel of Darwinism. It is precisely the weak spot in the Cambrian controversy that Stephen Meyer addressed when he wrote the peer-reviewed paper published in the *Proceedings of the Biological Society of Washington* in 2004. It is, in a sense, the ultimate question in the origin of life itself. DNA—its origin and diversification—is the ultimate issue of issues—the mother of all biological questions. And to that topic we now turn.

8

The Stubborn Mystery

How Did Life Begin?

When I was a high school student growing up in a farm town near Columbus, Ohio,[1] my brothers and I loyally performed a Friday night ritual. After coming home from a football or basketball game, we eagerly tuned in to a local television station at 11:00 p.m. to watch *Chiller Theater*, a weekly double feature that majored in horror and science fiction genres. Some of the most memorable offerings, besides my favorites, *The Forbidden Planet* and *House on Haunted Hill*, were the Dracula and Frankenstein movies. I will never forget the first time I saw the original (and perhaps finest) of these flicks, the 1931 Universal production of *Frankenstein*, starring Boris Karloff as Frankenstein's monster. Anyone who has seen this version, or one of the latter-day adaptations, surely will have riveted in his or her memory the image of the lightning storm zapping life-giving energy into the monster's limp body.

In the 1931 movie, after the decisive lightning bolt has surged life into the monster, Frankenstein, a scientist obsessed with the idea of creating life, triumphantly proclaims: "Now I know how it feels to be God!" This act of blasphemy not only removes him from his family, his friends, and his bride (in the story line), it actually caused a number of film censors across the U.S. to remove that line from the movie as it began to appear in theaters in

1931. Now, you may ask, what in the world does this scene have to do with the origin of life on Earth?

Stanley Miller's Breakthrough

Frankenstein's monster being zapped to life has been compared to the scientific image of *lightning jump-starting cellular life*, a powerful icon that has been etched onto the consciousness of millions of biology students across the world as they beheld in textbooks the spark apparatus of Stanley Miller. A graduate student in chemistry at the University of Chicago in late 1952, Miller set out, under the encouragement and guidance of Nobel laureate Harold Urey, to experimentally confirm a theory on the origin of life proposed in the 1920s by two scientists: Aleksandr Oparin in Russia and J. B. S. Haldane in England. Unlike Frankenstein, Stanley Miller's electric jolt of simple gases in a glass container wasn't intended to enable him to know what it feels like to be God. He was confirming that nature had the raw materials and talents (as Oparin and Haldane suggested) to mimic God—creating the building blocks of life by its own natural processes.

Stanley Miller and Harold Urey were pioneers in a field that now connects the work of several hundred scientists, working in *abiotic chemistry*—a field also called "abiogenesis," "prebiotic evolution," or "chemical evolution." (I prefer the final name.) Some researchers are chemists or biochemists; others come from disciplines such as geology, astronomy, and biophysics. Investigators keep in touch with each other's work through a loose-knit network known as the ISSOL—the International Society for the Study of the Origin of Life, which sponsors a newsletter and journal and holds a major conference every three years.[2]

Besides the common target of study, the workers in the field of chemical evolution share a powerful philosophical framework: they start by *assuming* that nature must have produced life on its own. No other option is on the table. The range of possible explanations is constricted to scientific law or chance (or some combination). Scientists working in this naturalistic tradition, which has hardened like concrete since Darwin, simply want to track down the most likely pathway, or at least one or more *plausible pathways*, that nature might have used. Phillip Johnson and ID theorists have repeatedly shown that in such thinking, one never asks first *whether nature could produce life unassisted by intelligence*. That nature did so is taken for granted. Rather, the key question is: "*How* did nature do it?" ID's critiques of chemical evolution raise questions about this assumptive base.

The scientists who inspired Stanley Miller—Oparin and Haldane—were building on that base as they proposed that a mixture of water vapor, ammonia, and other simple gases, energized by ultraviolet light or other sources of energy, could produce the basic structures of life. That is the idea Miller set out to test.[3] To set up a simulation experiment, Miller linked glass tubes to form a complete circuit and then fed into this loop a mixture of hydrogen, methane, ammonia, and water vapor, the very gases Oparin had proposed as life's raw materials. These were heated at one point in the loop and continuously circulated. Simulated lightning was added by means of a globe fitted into the pathway, into which two electrodes protruded. Every few seconds a spark would jump across the gap—from one electrode to the other through the hot flow of gases. After days of swirling and jolting the gases with the wimpy lightning flash, the mixture gave birth. Stanley Miller was able to draw out of the water (now a deep red) crucial samples, in which he detected several of the small molecules, *amino acids*, that are linked in living cells to form proteins. The results were clear by the end of the 1952 Christmas break at the University of Chicago. Miller joyfully had received his ultimate holiday present: *amino acids—birthed by sparks of electricity*.

What's the big deal about amino acids? For Darwinists, a naturalistic version of the "Hallelujah Chorus" was practically breaking out in 1953 as Stanley Miller's results were published, since amino acids were known to be proteins' all-important building blocks. Because life is unthinkable without proteins (see the end of chapter 5), Miller's research into pathways by which nature was said to have assembled the twenty building blocks of proteins without intelligent guidance was seen as a gigantic leap forward. His idea was that in the right environment—for instance, in an evaporating pond—these amino acids then could link together on their own to form a protein, which could combine with other organic molecules and eventually become encapsulated as a cell.[4] This image of a puddle of chemical broth enriched with such naturally growing protein chains quickly gave rise to the phrase "prebiotic soup" as the venue for Miller-type chemical evolution.

It's true that Stanley Miller's original experiment did not deal with the origin of DNA and RNA, the two other vital informational molecules, with their various building blocks. In his defense, we should keep in mind that the structure of DNA was only unraveled in 1953—the same year Miller's results were reported. He also did not investigate the origin of the lipid membrane—the cell wall so vital to life, even in humble bacteria. So Miller's experiment did not tackle all the parts of a cell. It was just a start, focused on the precursors of proteins alone. Nevertheless, it stands on its own as a significant event in the history of science, and that's why he and Urey became justly famous to the generation of scientists and students in the 1950s and in the decades since then.

Urey-Miller in Historical Perspective

For over fifty years Stanley Miller has continued to do research on the same kinds of problems—deriving biology's building blocks through processes thought plausible for the Earth's earliest history. As evidence accumulated that threw into question his preferred reducing atmosphere—the original mixture of gases with no free oxygen and plenty of hydrogen—he adjusted his experiments and found that they yielded much less impressive results.[5]

Other problems became manifest as well, such as the inability of the spark experiment to produce purely left-handed versions of amino acids that are needed for proteins.[6] Each of the amino acids used to build proteins, with one exception,[7] is naturally produced in two different mirror image forms in such experiments—in a right-handed form and in a left-handed form. Again these are stereo images of each other, much as a right-hand glove is the stereo or mirror image of its left-hand partner. Here's the odd fact: *Proteins use only the left-hand type of building blocks, also called "laevorotary" or "L amino acids," whereas the products of Miller's spark experiment were mixed left and right (such a mixture is called "racemic").* Spark experiments and the like always produce a roughly fifty-fifty mixture of both L amino acids and also the right-handed variety, or D amino acids (D stands for "dextrorotary" or right-handed).

This problem of a left and right mixture, with both D and L amino acids swimming in the solution, became increasingly evident and seemed to resist all attempts to solve it, even after decades of intense research. For example, in forming short proteinlike chains called "polypeptides," any L amino acid has been found to have no preference to link with another L type. In other words, L-with-D (left and right) linkages are just as chemically likely as L-with-L linkages. With no known mechanism in a simple pondlike environment to sort out and reject the right-handed amino acids and retain only the lefties, how could a true protein with left-handed amino acids ever be strung together in the first place? This problem continues to dog the entire field of chemical evolution to this day.[8]

So many problems with the prebiotic soup hypothesis became apparent (others will become clear below) that the importance of Stanley Miller's early work steadily receded into the background. In addition, a sense of priority of research increasingly led other workers to concentrate efforts elsewhere. While grateful for Miller's role as the one who launched their field's experimental phase, many new workers focused on the later stages, which are much more difficult stages for all scenarios of life's origin. They asked: *How can we explain the progressive self-assembly of the simple building blocks into the first informational chains (of DNA, RNA, and proteins) and then finally into a functioning replicating and catalytic system, at which point true life processes could kick in?*

Once these sketchy evolutionary processes have formed a primitive replicator, which can also perform the trick of processing energy from its environment, we have essentially arrived at what's called the "progenote,"[9] and it could then be truly described as *alive*. At that endpoint (a bit more primitive than ancient bacteria), the progenote is then ready to be acted on by natural selection, thus opening the door to Darwinian evolution. Yet getting to the progenote from a pool enriched with amino acids and other chemicals demands countless difficult and improbable steps across a vast and murky chasm.

As I move into the heart of our discussion of chemical evolution, I recognize the challenge of writing about this field, given the vastness (and technical difficulty) of the published material on chemical evolution. This has long been an area of special focus for me, but I sensed the need to dig deeper into the literature to write on the origin of life. I devoured articles and perused important new books. The more I read, the more I felt the enormity of the challenge. This material could barely be covered well in an entire book, let alone in two chapters! So I will proceed by describing the intense discussions and disagreements between ID and the leading figures in chemical evolution. *The Design Movement looks at this topic as a massive plus*—an area of evidence that is increasingly tilting in favor of Intelligent Design. Darwinists, on the other hand, are in a tough spot and are playing an increasingly defensive game as scientists learn more about this labyrinthine puzzle. Evolutionists paint the rosiest picture possible of a field that seems stagnant or enmeshed in stalemate.

The rest of our investigation is ordered in three steps. First, I will conduct a high-speed survey of the ups and downs of this field since Stanley Miller—the relevant discoveries and setbacks, the new ideas and their critical analyses. This will occupy the remainder of this chapter. Second, in chapter 9, we'll touch on new discoveries about the minimal DNA for a cell and will profile important "aha" moments for two scholars. Third, I will conclude by describing three responses to the question, "After all these years, are we making good progress in solving this mystery, and what are we learning about life?"

Mystery's Comprehensive Critique

In 1984, a trio of scientists published a polite but scientifically ruthless critique of the entire field of chemical evolution: *The Mystery of Life's Origin* (hereafter referred to as *Mystery*).[10] The book's cover sported enthusiastic blurbs from two scientists (NYU chemist Robert Shapiro and Dartmouth physicist Robert Jastrow) who had written on the topic of the origin of life from a naturalistic perspective but nevertheless viewed *Mystery* as a much-needed reality check. *Mystery*'s three authors were Walter Bradley, a materials scientist, Roger Olsen, a geochemist, and Charles Thaxton, a chemist and historian of science. Two of the three, Bradley and Thaxton, vigorously pursued this topic in the years that followed, and they emerged as leading intellectual architects of Intelligent Design from the mid-1980s onward. Thus the chemical evolution quandary was one side of the double womb from which ID was birthed (the other side was Michael Denton's *Evolution: A Theory in Crisis*). Since *Mystery* itself constitutes a multifaceted report of the field of chemical evolution in 1984, it makes sense to start our tour with six key highlights from this comprehensive critique of prebiotic evolution.[11]

- *Foreword to the book by biologist Dean Kenyon.* This was a surprise to some because Kenyon, a professor at San Francisco State University, was coauthor of an early leading chemical evolution book, *Biochemical Predestination*.[12] Here in *Mystery*, he made public his move toward skepticism of his theory in light of further research and rethinking.
- *The plausibility of DNA and RNA scenarios of random assembly was heavily critiqued.* Since DNA depends on proteins for its functioning, and yet proteins depend upon DNA (and RNA) for their own assembly, this is the ultimate chicken-egg question as to which came first. *Mystery* pointed out that just as proteins depend on left-handed-only amino acids for their construction, so also DNA and RNA require right-handed ribose or deoxyribose sugars. Researchers did experiments trying to link the building blocks of DNA (sugars, phosphates, and the four chemical letters). The result was short, poorly shaped chains. This reflected a DNA nightmare: if it is to form its lovely double helix shape, building blocks must be linked in a delicately precise way. Predictably, this result was never achieved in the rough and tumble of a raw, unstructured pond environment. Nature was proving itself incapable of knitting together even short DNA ladder segments.
- *"The Myth of the Prebiotic Soup"*—a chapter praised by leading researchers—detailed the constant threat of chemical ruination or termination of any growing chain. These unavoidable chemical glitches are called "interfering cross-reactions." Such nasty reactions are virtually inevitable in a pond, lake, or ocean where biological building blocks were said to be linking together. For example, if a protein chain or DNA ladder started to form, it would inevitably link with chemical

units like aldehydes that act to terminate the chain, sealing the chain's end and shutting down growth. Growing molecular chains would have to fight terrible odds to avoid such terminator reactions with chemicals swimming in any prebiotic soup.

- *An incredible number of oddball, nonprotein amino acids* (beyond the familiar twenty) was found in the mixtures of chemicals produced in experiments. These useless amino acids would inevitably join a growing chain, yet this would ruin the chain as soon as it joined. There seemed to be no mechanism to sort out the oddballs and keep them away.
- *Geological evidence seemed to turn against the Miller-type atmosphere.* Oxides deep in the Earth's rock record suggested the early presence of oxygen on Earth, and natural mechanisms would have created oxygen from atmospheric water.[13] On the other hand, there was no evidence in ancient rocks anywhere of a prebiotic soup.
- *Three chapters on the "thermodynamic barrier"* to forming life from a raw collection of chemicals were the most advanced section of *Mystery*, but this part won high marks from many reviewers.

Mystery pointed out that forming DNA or proteins can be compared with the assembly of words from Scrabble pieces thrown onto a table. The letters will have a disorganized, disorderly state—called a "high entropy" state. To lower the entropy (lower the disorder) by carefully ordering the Scrabble pieces (or amino acids) into specific meaningful words, it isn't enough just to have an energy flow. Energy flow simply won't help, because raw energy simply doesn't have any organizing capacity. To get the needed information, we have to configure the letters; we must specify the precise sequence. Raw, unharnessed energy flow cannot configure letters. This applies equally to Scrabble pieces on a table or to the amino acids swimming in the pond.[14] To do such configurational work, it takes two things that chemical evolutionists don't have:

1. You need a *mechanism to convert the raw energy flow into useful forms of energy*. For example, a garden can harness raw sunlight and grow corn and tomatoes, and those energy-rich molecules are consumed by hungry Scrabble players, whose muscles then turn the bio-energy into muscular kinetic energy as they move the Scrabble pieces.
2. You also need a *blueprint* to organize the letters in proper, specific patterns. An intelligent agent is implied, suggested *Mystery*, but the authors only mentioned this possibility in their philosophical epilogue along with other alternatives to the conventional approach.[15]

This concludes my brief sketch of several high points of *Mystery*. The positive case for design of the first cell, which was only hinted at in the chapters on thermodynamics and the afterword, was much more thoroughly developed in the two decades since *Mystery*, primarily by Stephen Meyer. He articulated the all-important role of the empirically observed structure of our world—that informational complexity is habitually and *exclusively* seen arising from intelligent agents. This concept—the uniform cause-effect structures of our cosmos—provides a key to what is called "the inference to the best explanation," which leads the investigator to the conclusion of design.

Hope in the Nineties: New Worlds and Proliferating Ideas

Mystery enjoyed a modest ripple effect among scientists in general, and especially in the chemical evolution field. In the *Yale Journal of Biology and Medicine*, medical professor James Jekel wrote, "The volume as a whole is devastating to the relaxed acceptance of current theories of abiogenesis." Klaus Dose, a leading prebiotic researcher, referred favorably to *Mystery* in a 1988 review article and summed up the situation: "More than 30 years of experimentation on the origin of life in the fields of chemical and molecular evolution have led to a better perception of the immensity of the problem of the origin of life on Earth rather than to its solution. At present all discussions on principal theories and experiments in the field either end in stalemate or in a confession of ignorance."[16]

Stagnant and *stymied* might be two words that in 1988 described the sense of struggle in elucidating the pathway from lifeless chemicals to the living cell. Yet hope seemed to revive in the late 1980s just as Dose was summing up the grim situation. New theoretical worlds were glimpsed, especially the *RNA world*. This vision was based on the new "RNA-first" idea, which sprang from research by a pair of scientists who shared the Nobel Prize in Chemistry in 1989: Sidney Altman of Yale University and Thomas Cech of the University of Colorado. In the 1980s they discovered that sometimes RNA can mimic certain proteins, showing a possible way out of the chicken-egg question (which came first, proteins or DNA?). In effect, they said, "Neither came first—RNA came first!" The key to this idea is their discovery that some RNA molecules *can act as enzymes*, causing certain chemical reactions to speed up, much as today's enzymes (specialized proteins) do constantly in our cells. Thus, RNA molecules can be seen to perform double duty: they store information (like DNA) and yet, surprisingly, they sometimes catalyze chemical reactions. The best of both worlds can be had in one molecule!

RNA went overnight from the forgotten molecule in the shadows to the new star of origin-of-life studies.

However, a new set of questions emerged: Where did the informational sequences in the hypothesized RNA proto-life come from? (This was a key question raised over and over by leading figures in abiotic chemistry.) How did the crucial right-handed-only ribose sugars get sequestered and linked into this molecule from a soupy mixture containing both left- and right-handed sugars? Is there evidence that RNA can copy itself? The answers to these questions have not been helpful. Robert Shapiro, a nontheist who applauded *Mystery* in his dust jacket blurb and who then wrote his own celebrated review of chemical evolution,[17] summed up the RNA scenario's plight in 2000: "A profound difficulty exists, however, with the idea of RNA, or any other replicator, at the start of life. Existing replicators can serve as templates for the synthesis of additional copies of themselves, but this device cannot be used for the preparation of the very first such molecule, which must arise spontaneously from an unorganized mixture. The formation of an information-bearing [RNA chain or equivalent] through undirected chemical synthesis appears very improbable."[18] We are back at the problem of configuring the first chemical Scrabble pieces!

In spite of such criticisms, chemical evolutionists worked on the RNA world as the best idea for the early steps before DNA and proteins appeared. Yet beyond the hype of the RNA world, from 1990 through the present, a dizzying number of new ideas were proposed. I'll just list a half dozen of these so that the reader can get a feel for the intense activity in this period:

- *Christian de Duve, Nobel laureate in cell biology, entered the field.* He launched a second research career focused on chemical evolution after his retirement from Rockefeller University in 1988.[19] His work culminated in *Vital Dust*, an exploration of the origin of life. He is known for having congenially interacted with a number of ID researchers at the end of the 1990s and beyond, even speaking at a conference hosted by William Dembski. De Duve proposed a forerunner to the RNA world, known as the "thioester world."[20]
- *Announcing: The iron-sulfur world.* This was proposed by German attorney and chemist Gunter Wächtershäuser (pronounced "vock-ter-zoy-zer"), who envisioned life beginning with the help of iron pyrite—known as "fool's gold." According to this view, the ideal venue for the critical buildup of the early prebiotic compounds on fool's gold was deep-sea vents, where key reactions would be likely to produce the iron pyrites.[21]

- *Graham Cairns-Smith hatched a bizarre idea.* He suggested that life began as a mineral-based entity, a sort of "clay crystal life," and then after a lengthy period of time, an abrupt transition took place when the mineral world gave way to a biochemical successor. This is called the "genetic takeover."[22] In addition to Cairns-Smith's radical idea, minerals in general were deemed such a promising structure to assist as midwives in life's origin that one chemical evolution text has a chapter that is subtitled: "Minerals Functioning as Scaffolds, Adsorbents, Catalysts, and Information Carriers." One popular idea was that a special clay called montmorillonite served as a template where primitive chemical units of information might be linked in organized arrays.
- *Chemical evolution in outer space?* This newer side of prebiotic evolution studies experienced a surge of interest in the 1990s, especially focused on Mars and on the moons of Jupiter and Saturn. This focus on outer space, encouraged by grants from NASA's exobiology program, also paid attention to the idea of amino acids and other building blocks of life being delivered to Earth aboard meteorites and comets. Interest seemed to wax and wane periodically in regard to the notion of panspermia (spores of life floating to Earth from elsewhere) or even directed panspermia (simple life-forms being sent to Earth in a spaceship). These seemingly science-fictional ideas of life drifting or being sent from some other place in the cosmos, though they were generally considered unworthy of serious consideration before 1980, were popularized somewhat by Francis Crick's flirtation with the idea in *Life Itself.*[23] Such published speculations indicate that chemical evolution on Earth has come to be viewed by some researchers as exceedingly implausible.
- *Behold: bizarre bacteria.* A wave of excitement surged over weird kinds of bacteria, called "Archaea," discovered by Carl Woese and others. These bacteria, also called "extremophiles," seemed to have in common a love for harsh settings, such as extreme pressure, incredible heat or cold, and normally lethal acidity (or alkalinity). Unknown before the 1970s, they were quickly hailed as the probable survivors from the earliest life-forms, when Earth experienced harsher environments. Their DNA sequences were strikingly different from normal bacteria.
- *How much time for evolution?* Finally, the amount of time envisioned for chemical evolution to reach its goal kept getting smaller and smaller, until the window had shrunk to a remarkably small size. (Some said it could have happened as fast as in mere tens or hundreds of years, sandwiched in between devastating meteor impacts on the early, hell-

ish surface of the Earth!) William Schopf, a pioneer in such studies, said that fossilized remnants of fairly sophisticated bacteria seem to appear in rocks dated 3.5 billion years old. Further studies in rocks older than 3.8 billion years reveal telltale traces of carbon isotopes, indicating the likelihood of biological activity. Accepting these dates as accurate for the sake of discussion, and assuming that at 4.0 billion years ago the planet was far too hot, molten, and hostile a place to allow life to survive, ID theorists pointed out that the extended "window of chemical evolution" had been squeezed down to an extremely tiny fraction of the available time that was previously assumed.

To sum up, the field of chemical evolution in the 1990s was bubbling up with new ideas, strange discoveries, and—very significantly—a shrinking time frame in which all of the intricate complexity and biochemical information of a single cell could have been produced from simple chemicals. There was no lack of imaginative stories. Chemical precursors were zapped, or heated, or rooted on clay surfaces, or irradiated into existence in a pond, or by the side of bubbling ocean vents, or in the cold stretches of outer space. The field seemed to have no lack of colorful venues for the evolution of life to take place. *But what it seemed to desperately lack—and this became more evident and stark over time—was a plausible mechanism to assemble the vital chains of DNA, RNA, and proteins, and then to organize them so as to form the first living cell.* Yet that question itself depended on another question: How genetically simple can one get and still have a functioning cell? That is the key question that had to wait for genome sequencing techniques developed in the late 1990s, and that is our next story.

Assessing the Origin-of-Life Question

What Have We Learned?

In August 2005 I emceed a conference entitled "Uncommon Dissent: Scientists Who Find Darwinism Unconvincing." This three-day gathering, held at a cavernous convention center in Greenville, South Carolina, focused on many scientific fields where empirical evidence had come to favor an Intelligent Design explanation over a Darwinian one. Excitement ran high in the meetings because just three days before the Thursday evening kickoff, President Bush had sent the media into a frenzy when he revealed that he thought it was a good idea for students to be exposed to ID theory. Timothy Chu, a *Time* science writer, covered our event. He circulated among the speakers, including three ID superstars: Michael Behe, Paul Nelson, and Jonathan Wells. Thus it was no surprise when several key quotes in the following week's *Time* cover story came from our conference speakers.

When I first scanned the list of speakers in late spring, I was surprised that half were new names. Most of these were scientists teaching at prestigious private institutions or large state universities. These new names were a clear plus; they meant that new blood was rising, that biologists and chemists at universities were doing research on design and sharing their findings. It was a scientific feast, affording a good glimpse of the research side of ID. Speakers showed why they are increasingly convinced of design as they work in the trenches of experimental research.

Of eight superb presentations, two focused on evolutionary scenarios leading to the first cell—oceanographer Edward Peltzer and David Keller, a biochemist at the University of New Mexico. One key issue was tackled by David Keller and his colleagues: *How complex is the minimally conceived living cell?* What kind of gap of biological information are we looking at between (a) the hypothesized RNA world ancestor (for instance, a single replicating RNA string, equivalent to one gene of DNA), and on the other hand (b) the simplest living cell? To answer this question, scientists need to determine what is the minimum quantity of genetic information at the threshold where an entity can barely run and replicate itself—and thus deserve to be called a cell.

Years ago a famous evolutionist told me that a cell could probably make it with just forty genes—each gene having three hundred to several thousand letters. As it turns out, this professor was not even close; his guess was *way too low*. By the late nineties, scientists had answered this huge question, using a variety of techniques including new genome sequencing methods made famous by the Human Genome Project.[1] Published estimates began to appear for the minimum threshold of information to operate just the simplest *independent bacterium*. By "independent," scientists mean a cell that can make it on its own, without being coddled and spoon-fed by other associated cells or tissues. Even though humble E. coli (a species of bacteria swimming in our gut) possesses an astonishing number of genes—4,288—scientists tracked down some slimmed-down bacteria that can survive on their own with a much smaller set, in the range of 1500–1900 genes.[2]

But is this the lower limit? Scientists were unsure. They pressed further, peering inside the genome libraries of the simplest organisms on planet Earth, especially the parasitic microbes such as Mycoplasma genitalium. Such critters, with gene totals in the range of 470 to 863, rely on a steady supply of key materials (sugars, fatty acids, amino acids) flowing from the hosts they depend on. These are not hardy, independent microbes; they're no genetic Rambo. On the contrary, they are so feeble and DNA-challenged that they practically "loaf by the poolside of a rich uncle" (the host organism), being fed steady trays of food by butlers, who wait on them to keep them alive.[3] Various estimates of these lazy organisms' minimum gene package, which surely is a safe lower limit for any *independently alive single-celled entity*, were published in recent years. Sometimes the estimate was about 400 genes (the extreme parasite Buchnera, for example, needed a minimum gene set of 396); others were estimated to be able to survive with a bit smaller set of genes. One study with a Mycoplasma "determined the minimum number of genes to fall between 265 and 350," while another study of a bacterium (Bacillus subtilis) "estimated the minimal gene set numbers between 254 and 450."

The lowest estimate published anywhere was 246 genes, so it is safe to say that no organism can exist with less than about 250 genes. Although this seems to be the lowest possible limit, it is rather dubious that the lowest genetic package for life could be that low. Independent cells may actually need, when the truth is ultimately known, much closer to the higher level (1500–1900 genes), which are actually known for independent (nonloafing, nonparasitic) super-simple bacteria.[4]

The conferees in Greenville were reminded that to get a functioning cell, we don't just need the DNA package in place. Cells also need the materials of the entire interior space carefully arranged in their 3-D positions, complete with the all-important ribosome machine that reads the recipes inscribed in RNA and then builds the correct proteins. Also the cell wall has to be accounted for. The conclusion drawn by Edward Peltzer, David Keller, and others was that chemical evolution investigators had actually made the most interesting headway in explaining how the very first building blocks (amino acids, sugars, and so on) could be synthesized. (Both emphasized the implausible oxygen-free atmosphere used by Stanley Miller and others. Yields of building blocks using plausible atmospheres were much less impressive.)[5]

The later stages of chemical evolution (linking DNA, RNA, and proteins into meaningful chains for replication and metabolism; developing the cell wall) are another matter entirely. After fifty years of study, chemical evolution research has shown little if any progress in elucidating these later stages. The forming of integrated swarms of information-packed molecules seems increasingly mysterious, given our growing knowledge of the problems involved. The RNA world research, a focus of overwhelming experimental work recently, seems to be turning up more discouraging results each year (many problems were explained by Keller). In its current state, they argued that abiotic chemistry utterly lacks any plausible scenario for the assembling of the incredible amount of genetic information—now known to be a set of 250 genes minimum, but more likely over 1,000 genes—that is needed to run the simplest, most feeble cell one can conceive. Nor does any plausible scenario seem to be lurking on the horizon.

Prebiotic "Aha" Moments

As I listened to the two lectures, I marveled at how much has been learned about the issues relevant to life's origin since the Stanley Miller experiment. I also wondered: Are workers in this field opening up to the *mere consideration* of a role for intelligent agency in life's origin? Or is the core assumption of the naturalistic paradigm—"We know nature did it, period"—still accepted without hesitation? It seemed that in light of the new "minimum genetic package" studies, the tension between scientific reality and the paradigm's assumptions must be approaching the snapping point. One would think that restless minds might toy with the *theoretical possibility* of an intelligent cause. There are, in fact, two golden examples of scientists who had such heretical thoughts.

Case #1: Dean Kenyon

Unlocking the Mystery of Life (hereafter referred to as *Unlocking*), which I discussed in chapter 4 and elsewhere, has a fifteen-minute section that features some extremely well-done video animation of DNA, RNA, and proteins in action. These video graphics are woven into a concise recounting of the change of mind of prebiotic evolution researcher Dean Kenyon, who teaches biology at San Francisco State University. I related his change of mind briefly in chapter 8, in citing his foreword to *The Mystery of Life's Origin*.

The story is well told in the video; here I'll just mention some highlights that underline Kenyon's *empirical* motivation to rethink his views. The evidence shows that he was willing to reconsider his theory only in light of the totality of biochemical evidence that he was tracking, and only after he studied a scientific critique that had been published. Originally, Kenyon felt that the natural bonding preferences of amino acids would explain the rise of certain preordained types of chains, containing ideal patterns of amino acids that were predestined by chemistry. From this key idea came Kenyon's book title: *Biochemical Predestination*. His book says, "Life might have been biochemically predestined by the properties of attraction that exist between its chemical parts—particularly between amino acids in proteins."[6]

However, within a few years after its publication, Kenyon encountered a published critique of his position given to him by one of his students. Over summer break he read and tried to refute this counterargument. In the end, he found that he could not, and he began to rethink his former position. One by one, options seem to be closed off. Chance alone, Kenyon realized, could never compile the quantities of information needed to code for all of the proteins in a single-celled animal. Natural selection also seemed like no help, since the replication process (needed for selection) apparently depended upon the presence of the genetic material in DNA, and yet it was the origin of DNA that needed to be explained. Biochemical predestination seemed impotent as well. Life's encoded messages don't seem explicable, after all, as the result of lawlike forces, even working with random shuffling processes. Law and chance, even working together, seemed incapable of producing the digital-type information sequences in DNA and proteins. In an interview featured in the *Unlocking* video, Kenyon recalls this intense period of thought:

> It's an enormous problem how you could get together, in one tiny submicroscopic volume of the primitive ocean, all of the hundreds of different molecular components you would need in order for a self-replicating cycle to be established. And so my doubts about whether amino acids could order themselves into meaningful biological sequences on their own, without preexisting genetic material being present, just reached for me the intellectual breaking point near the end of the decade of the 70s.[7]

Kenyon also was impressed by experimental results he was getting: "The more I conducted my own studies (including a period of time at the NASA-Ames Research Center), the more it became apparent that there were multiple difficulties with the chemical evolution account. And, further experimental work showed that amino acids do not have the ability to order themselves into any biologically meaningful sequences." The key for Kenyon's abandonment of naturalistic scenarios of the origin of life seems to have been a focus on the origin of the genetic material in DNA: "If one could get at the origin of the messages, the encoded messages within the living machinery [of a cell], then you would really be onto something far more intellectually satisfying than this chemical evolution theory."[8] What made Intelligent Design immensely attractive to Kenyon as a scientist is that it accorded well with the universal causal pattern seen in our experience of the universe—namely, that dense concentrations of digital information are always seen to arise from intelligent agents and never are seen arising from natural or unintelligent processes like chance and scientific law.

One of the most powerful moments of the *Unlocking* documentary takes place when DNA is pictured being opened to let RNA form, which then zips out of the nucleus and attaches to the "reading machine"—a ribosome. This machine then sequentially reads the RNA as a template for a growing protein. At the end of this two-minute sequence, Kenyon expresses his amazement at the level of technology here, along with its implications: "This is absolutely mind boggling—to perceive at this scale of size such a finely tuned apparatus, a device that bears the marks of intelligent design and manufacture. And we have the details of an immensely complex molecular realm of genetic information processing. And it's exactly this new realm of molecular genetics where we see the most compelling evidence of design on the earth."[9] Clearly, Kenyon's own intellectual epiphany was initiated through an honest attempt to rebut a scientific argument against his view, and it came to completion as he pondered the most likely source for the cell's information processing system.

Case #2: Paul Davies

If I had to choose one book for every student to buy and read on chemical evolution, it would come down to a virtual tie—Fazale Rana and Hugh Ross's powerful *Origins of Life* (2004) comes in first by a nose, and Paul Davies's older (1999) but eloquent *The Fifth Miracle* comes in second. What is unique about Davies is his rather striking epiphany that, while different from Kenyon's, shares a number of driving concerns and some fruitful implications.

Davies is an Australian mathematical physicist whose prodigious output of books over the past several decades has been devoted to explaining the eerie fine-tuning of the cosmos. Yet he is not a part of the Intelligent Design Movement. In his books, he seems to favor a deistic conclusion that might be drawn from the study of cosmology and the physics of fine-tuning, but he balks at suggesting a full-blown intelligence who can intervene in the cosmos today. On this point, Phillip Johnson described a tense interaction he witnessed at a science conference between Davies and Christian de Duve, while the two were discussing the very points Davies brings out in *The Fifth Miracle*. In brief, Davies was making some comments in his talk that seemed to flirt with design in the origin of life. At lunch, Johnson witnessed a cross-examination by de Duve, which elicited a disavowal by Davies of any such heresy. The danger had passed, and de Duve seemed satisfied that Davies's hints of dark irrational thoughts had been squelched.[10]

Davies seems to abide by the unwritten rule of naturalistic science: "Law or chance or their combination are the only allowable types of explanation. Intelligent agency, within the cosmos at least, is disallowed as an explanation, since it moves us outside of science." Such an assumptive base (critiqued previously) is clear in *The Fifth Miracle*: "Of all the complex structures produced by terrestrial biology, none is more significant than the brain, the most complex organ of all. Are brains just random accidents of evolution, or are they the inevitable by-products of a lawlike complexifying process?"[11] To Davies, everything in the universe is explicable by scientific law or chance. (At this point, I see C. S. Lewis raise an eyebrow and ask, "Does that include Davies's own speculations and conclusions?")[12]

On the other hand, Davies makes it clear throughout the book (see pages 90–93 especially) that ordinary, well-known laws of nature simply will not—*they cannot*—do the work of building the informational sequences in the cell that are absolutely necessary for the origin of life. To say that Davies "gets it"—that he glimpses the problem of information in the context of life's origin—is an understatement. In his closing chapter, "A Bio-Friendly Universe?" this profound unease and subtle protest against the predominant thinking in the origin-of-life field reaches a climax. He says that it may be

> that life does indeed reside beneath Europa's icy skin, either for the relatively trivial reason that it traveled there from Earth in a meteorite, or for the much more profound reason that life is inevitable given the right conditions. According to the deterministic school of biology, which seems to dictate the prevailing view at NASA and is shared by most media commentators, life will automatically form in any Earth-like environment. Take a measure of water, add amino acids and a few other substances, simmer for a few million years, and—hey presto!—it lives. This popular theme is sharply criticized by the opposing school, which stresses the awesome molecular complexity of even the simplest living thing. To proponents of the latter position, the sheer intricacy of life bespeaks a freakish concatenation of events, unique in the cosmos. No amount of water, they say, even if laced with fancy chemicals, will come alive on cue. Earthlife must therefore be a fluke of astronomical improbability.[13]

Clearly, Davies is mocking the glibness of NASA and media commentators. He greatly appreciates the improbable result of arriving at cellular complexity. Yet he seems caught between a scientific rock and a hard place. What is the overriding concern for Davies in deciding which pole to lean toward? It seems elsewhere that he is attracted to Stuart Kauffman's famous "complexity theory," but even with Kauffman, he remains adamant that something is missing:

> Kauffman doesn't claim there is a pre-existing blueprint for life, or a propensity for organized complexity to emerge under suitable conditions. So life may not be such a surprise after all. . . . According to Kauffman's theory, there is no specific end goal encoded in the principles of self-organization, . . . only a general trend towards the sort of complex states that are likely to lead to life.
>
> Attractive though these arguments may be, we are still left with the mystery of where biological information comes from. . . . If the normal laws of physics can't inject information, and if we are ruling out miracles, then how can life be predetermined as inevitable rather than a freak accident? How is it possible to generate random complexity and specificity together in a lawlike manner?[14]

So Davies seems to come right up to the brink of a clear view of the law-versus-chance dilemma, and yet he clearly rules out *intelligence* by wrapping it in the scare word, *miracles*. In this final chapter on the one hand, he shrugs his shoulders on the choice between law and freak accident (at the end, one feels suspended between the two poles). Yet on the other hand, one can see his preference: new, hitherto unglimpsed *complexifying laws of nature* that can mimic intelligence.

> My own opinion is that emergent laws of complexity offer reasonable hope for a better understanding not only of biogenesis, but of biological evolution too. Such laws might differ from the familiar laws of physics in a fundamental and important respect. Whereas the laws of physics merely shuffle information around, *a complexity law might actually create information, or at least wrest it from the environment and etch it onto a material structure*. . . . My proposal means accepting that information is a genuine physical quantity that can be traded by "informational forces" in the same way that matter can be moved around by physical forces.[15]

In summary, Davies sees clearly the same intense information problem that Kenyon does, but he dodges the scientific pariah of "intelligent agency" by a clever move: he absorbs the function of intelligent selection into some yet-to-be-discovered "informational law." The enigma of the origin of information led Davies not to Intelligent Design per se but to a parallel idea rather more tame and infinitely more acceptable to his fellow scientists.

Three Perspectives on Progress (or Lack Thereof)

What the public is told (and especially what science students are taught) on chemical evolution hinges on the answers we give to two key questions: Is the origin-of-life question well in hand, or is the field deadlocked and stalled? What have we learned from all this scientific effort? Answers to these questions depend on who is speaking. I see three types of answers:

1. The Deadlock Dodgers

A minority of Darwinists will dodge the embarrassment of chemical evolution's deadlock, saying, "This field does not concern biological evolution by natural selection." Of course this is true, but such a response also seems irrelevant in view of the overall naturalistic project: explaining the origin of physical structures of the universe that appear designed, apart from intelligent agency. Stephen Jay Gould seemingly plays the role of a dodger in a remarkable footnote in his final book, *The Structure of Evolutionary Theory*.[16] Gould lists the points of a creationist syllogism (comprised of three premises and a conclusion), and then he goes on to *agree* with premises two and three: "(2) evolutionists can't resolve this issue [the ultimate origin of life—stated in premise one]; (3) the question is inherently religious." Gould agrees with these? Perhaps the reader finds this as astonishing as I do. Gould then expands on his comment in this footnote, saying that he and his fellow evolutionists "therefore do not study the question of ultimate origins or view this issue as part of scientific inquiry at all." He even quotes Darwin himself in defense of this way of thinking. Sometimes such dodging-Darwinists, after disconnecting themselves from the origin-of-life question, may add a note of cautious optimism—something like: "Nevertheless, we're making reasonably good progress in chemical evolution, since so many interesting ideas are now being worked on."[17]

2. The Doggedly Determined

More scientific commentators of a naturalistic bent will simply say that chemical evolution theorists are working as hard as they can on a massive problem, which is far from solution but for which (they remain confident) a solution can ultimately be found. This is well represented in a massive and remarkable book by Israeli chemical evolutionist Noam Lahav, called *Biogenesis: Theories of Life's Origin*. I could not resist ordering this book after hearing anti-ID author Niall Shanks point his readers to it. Shanks's own presentation on the origin of life in *God, the Devil, and Darwin* struck me as scientifically weak, so I thought Lahav, his favored authority, might present a more convincing case that science is nearing a solution in chemical evolution studies. I see why Shanks recommended Lahav; his work is brilliantly written, massively documented, and endlessly fascinating. But Lahav's own epilogue nicely sums up the discouraging state of affairs:

> Having made a long and tortuous journey in search of the origin of life, some readers may feel disappointed: The alarming number of speculations, models, theories, and controversies regarding every aspect of the origin of life seem to indicate that this scientific discipline is almost in a hopeless situation. Still others may find comfort in the significant progress already made and the knowledge accumulated in this inter-disciplinary discipline. . . . However, none of the theories advanced so far encompasses all the aspects of the emergence of the central functions of extant cells, thus "bridging the gap between life and inanimate matter" (Arrhenius et al., 1997). Furthermore, because some of these theories differ so widely from each other, bridging the gap between them seems difficult, perhaps even impossible, at present. Can we observe the initiation of promising directions that might lead us into the beginning of a new era in the study of the origin of life?[18]

In the next paragraph, he adds that "every researcher focuses on the potential of his or her own school to bring about a significant progress or even a breakthrough of a kind in the understanding of the origin of life. *It is difficult, however, and probably impossible, to point out which of the present schools of thought or their combination have the potential to serve as the basis for the next paradigm.*"[19]

I would encourage the reader here to simply reread these remarkable quotes above and reflect on what Lahav is saying. This honest and humble epilogue is so intellectually refreshing, it ought to be required reading in every high school or university class when studying chemical evolution. Note that he asks if there is any "initiation of promising directions that might lead us into the beginning of a new era"—a new paradigm—of the field in which he is an authority. In the wake of Lahav's wistful longing for a new era and new paradigm, he offers one suggestion. If scientific data in the near future bring a new certainty as to how much time was available for the evolution of the first living cell, then we may have a basis for choosing one approach over the others. Yet Lahav notes that "even if a paradigmatic change in the study of the origin of life took place and 'living' chemical entities were synthesized out of inorganic compounds in our laboratories, we should recognize our own limitation in deciphering the transition from inanimate to animate."[20] This welcome humility about the prospects of chemical evolution theory has occasionally spilled over (as in Davies's case) to the edge of acknowledging the role of intelligence.

3. The Design Detectives

The third kind of answer to "How's it going in prebiotic evolution?" comes from design theorists who have researched and written on the topic. They explain in detail the empirical problems and barriers that have been discovered over the years and point out that this increase in knowledge is a fantastic plus for science: it is precisely this new knowledge that keeps adding more weight to the conclusion that life had to be designed by intelligence. In other words, the inference to design would be much more tentative if we had not expended such efforts to check out every nook and cranny of the sprawling problem of life's origin. To conclude that life is designed is simply to follow the evidence.

This point of view is illustrated most recently by Walter Bradley's "Information, Entropy, and the Origin of Life," in *Debating Design*. Bradley's opening sections, "Quantifying the Information in Biopolymers" and "The Second Law of Thermodynamics and the Origin of Life," are fairly technical and forbiddingly mathematical for most nonscientists, but his last section, "Critique of Various Origin-of-Life Scenarios," is clear and blunt. This is where he scrutinizes competing theories, evaluating such proposals as Eigen's scenario of a simple self-replicating RNA, Kauffman's and Prigogine's models of self-organization, and the metabolism-first models of Wicken, Fox, and Dyson, which rely on *preferential bonding* between amino acids. In the latter case, Bradley explains the inherent problems with each approach, citing his own lab research, which falsified the preferential bonding hypothesis.[21]

Bradley concludes with a summary quote from a *New York Times* science review by Nicholas Wade: "The chemistry of the first life is a nightmare to explain. No one has yet developed a plausible explanation to show how the earliest chemicals of life—thought to be RNA—might have constructed themselves from the inorganic chemicals likely to have been around on early Earth. The spontaneous assembly of a small RNA molecule on the primitive Earth 'would have been a near miracle,' two experts in the subject helpfully declared last year."[22]

Bradley assessed the situation in 2004 as one in which the origin of life appeared to be the ultimate case of irreducible complexity. Thus, the more we learn about information and entropy, and the more we look at each of the alternative theoretical solutions for life's origin, the more confident we become that nature, unassisted by intelligence, is simply incapable of assembling on its own a living, replicating system. *This reality is the opposite of the "God of the Gaps" situation, where some gap in knowledge in science tempts a theist to use God as an explanation. The Intelligent Design conclusion in this field, says Bradley, isn't fueled by gaps in science but rather by the growth of science.*

Out of the Labyrinth . . .

To summarize, in recent years chemical evolution has been experiencing a sense of stalemate, leading some workers to wonder out loud if the field will ever arrive at a satisfactory solution. The tone of frustration we have heard from theorists and the sense of longing for a new paradigm is not uncommon in the field; sometimes the research results bubble up in open discouragement. Like Bradley, I view this stalemate as a wonderfully healthy product of the growth of science. It leads scientists to be open to radical new natural-cause ideas (e.g., Davies's information-creation laws), but it can also lead to a profound reassessment of the field's dominant approach (as in Kenyon's case), with its tightly held assumptions about ultimate reality.

A vivid underground metaphor keeps haunting my mind.[23] I picture scientists working in chemical evolution as explorers trying diligently (and without success) to work their way out of a huge underground labyrinth. Those who rule the labyrinth mapping committee—philosophical naturalists—are absolutely sure that the only way out is up a major tunnel system—labeled on their maps as the "natural cause tunnel system," with branches of all shapes and sizes being mapped out daily as explorers fan out into this system. As the scientists deployed in these tunnels bring back their reports, the mappers write down every bulge and indentation, every crooked, dead-end shaft.

Looking for that passageway out, they are always learning more, but some on the committee have begun to wonder if they are making real progress to the goal. The mapping is almost complete now, and a quiet discouragement is sinking in. Rarely do they verbalize whether they are working on the right tunnel system in the labyrinth—after all, what's the alternative? Oh yes, they know about the other main tunnel marked "intelligent cause," but how can that help? No, that's religion; it's not science!

This "natural cause tunnel system" is increasingly manifest as one huge and intricately complex network of dead ends when it comes to the origin of new genetic information. As scientific tunnel searchers try in vain to find a way that *unintelligent nature* can build a living cell from scratch through many intermediate steps, they are merely mapping, month by month in ever greater detail, the fine contours of an elaborate cul-de-sac.

Might some researchers eventually cast a glance at the other major tunnel system, heading off in a different direction—that maligned tunnel marked "intelligent cause"? Dean Kenyon did; Paul Davies seems to have gazed intently at the entrance and turned away. Nobel laureate Richard Smalley, just months before he died at the age of sixty-two in November 2005, affirmed that, after restudying the evidence, life indeed must have been created by an intelligence. He had entered the tunnel.[24]

Other scientists, dissatisfied with current answers yet biased philosophically against design, have paid little attention to design, but that may be changing. Word is spreading not only that chemical evolution is stymied but also that a carefully defined and mathematically coherent notion of information in biology—Complex Specified Information (CSI)—is now in place and is being vigorously discussed. Also, more scientists are hearing that the main developers of the CSI theory (William Dembski and Stephen Meyer) have also set up a principled system that relies on statistical analysis and careful testing of evidence, rather than hunches or leaps of intuition. This new theory is already tempting more and more scientists to cast a glance at the labyrinth's other forbidden tunnel system.

10

CSI and the Explanatory Filter

Dembski's Trial by Fire

"*Gort, Klaatu barada nikto*!" Trivia buffs treasure these words, uttered by actress Patricia Neal to a huge silver robot named Gort in the 1951 sci-fi classic, *The Day the Earth Stood Still*.[1] The movie plot hinges on Klaatu (played by British actor Michael Rennie), a galactic emissary who lands his flying saucer in the shadow of the Washington Monument to deliver an ultimatum to the peoples of the Earth: learn to live peacefully together or be destroyed as a danger to other planets. Klaatu is accompanied by Gort, the robot whose fearful laser, beaming from a slot on his head, vaporizes an army tank after a jumpy soldier shoots and wounds Klaatu. After being treated at a hospital, Klaatu slips away from the authorities detaining him and transforms himself into "Mr. Carpenter," a boarder in the home shared by a young widow, Mrs. Benson (played by Patricia Neal).

I'll spare you the details of the plot and jump to the chase scene, where Klaatu, sensing his impending capture, implores Mrs. Benson to memorize the strange words. He tells her that if anything happens to him, she should go and utter the words to Gort, who is immobilized outside the spaceship. Let me stop and ask two questions: (1) *How do we know that this cryptic phrase contains true information, as opposed to mere gibberish?* (2) *Can we know what the message means?* To show that the phrase is not gibberish (a meaningless jumble of sounds), all we have to do is show that it is likely a meaningful set of words in the context of the movie action. But asserting that it is "meaningful" simply raises the more specific question: what precisely do the words mean? We can only guess. It functions like an order to activate Gort, so perhaps it is a simple command, "Go rescue Klaatu!" Yet in Gort's computer brain, it is conceivable that it could mean much more, such as "Klaatu says initiate emergency sequence XV-6," where XV-6 is a set of instructions preprogrammed to unfold logically, including (1) tracking down Klaatu, (2) dealing with threats and barriers along the way, and (3) making all necessary follow-up decisions to help Klaatu. (If you've seen the movie, you'll know why I say this.) The bottom line is Gort's response; he is definitely activated by the command, and that is the clearest evidence that the phrase is indeed meaningful.

Does lexical analysis help pinpoint meaning? Leaving out "Gort," the message is typically written as three words with a total of seventeen letters, with one word, *Klaatu*, having a known meaning. The other two words are unknown; we seem to have come to a dead end. All we can say is this: the meaning supplied by the three words, using seventeen letters, appears to be a command designed to activate Gort on behalf of Klaatu. Yet in the final analysis, we don't have to know the exact meaning of "*Klaatu barada nikto*" to detect the presence of true *information*—a sequence of symbols that carries meaning and that displays the two crucial qualities of complexity and specification. This string of symbols is characterized first as *complex*—it has a pattern with a number of constituent parts, which don't just keep repeating, like the sodium and chloride atoms in a salt crystal. Second, it is *specified*—each symbol in the string is stipulated, or chosen, to some extent. The string cannot tolerate free substitution of verbal symbols (verbal mutations) if the meaning is to be retained. That's why Klaatu had Mrs. Benson repeat the phrase during the chase scene until it was accurately memorized with each vowel and consonant in place. So Klaatu's meaningful phrase displays specified complexity. It does indeed contain information.

What does all this have to do with Darwinism and the struggle over Intelligent Design? Practically everything, because *specified complexity* is defined by ID theorists as the universally reliable indicator (criterion) of an intelligent cause that has left its mark. Where you find segments of complex information that are highly specified, you have the fingerprints of intelligence. Note the movie's parallel with biology: Klaatu's three-word phrase, with its seventeen letters, is structurally identical to the language system of DNA and proteins. One of the shorter proteins, cytochrome C, has one hundred amino acids, essentially one hundred biochemical letters strung together to form the long protein word. But the amino acid sequence is the result of translation from the *cytochrome C gene*—a functional blueprint with a DNA-RNA sequence of slightly over three hundred genetic letters, which are arranged in one hundred codons (three-letter words used by DNA and RNA). The DNA-RNA parallel with human (or robotic) sentences is enhanced by the presence of special words at the two ends of gene sequences, the start codon and stop codon. These special genetic words function like the capital letter and period marking the start and end of a sentence.[2]

The inference from information to intelligence is further grounded on a vital yet often overlooked corollary, which is ID's main empirical claim regarding information: *We have learned that cells can shuffle information around and rearrange genes within the genome. But scientific observation and experimentation have failed to show that nature possesses the power to compose specified genetic information in significant quantities.*[3] Just as "*Klaatu barada nikto*" arose from intelligence and was communicated to an intelligently programmed, quasi-intelligent system (Gort's advanced computer brain), so ID argues that we can conclude that the segments of specified, complex information in the DNA, RNA, or proteins found in every plant or animal must have arisen from true design by some intelligence. That conclusion is intellectually rock solid, unless the day arrives in the future when we find out by experimental research that natural causes can indeed create specified complexity. One could even say that the *inference to design from specified complexity* is the crux of the matter being fiercely debated. It is ID in a nutshell, and its conclusions differ as radically as one can imagine from those of Darwinian theory.

Measuring DNA's Information

In posing the challenge of explaining the *origin* of biological information, Stephen Meyer and other ID theorists have fixed their focus on the information content of DNA with its four-letter alphabet, comprised of four nucleic acids: A (adenine), T (thymine), C (cytosine), and G (guanine). (RNA uses a similar set of letters but with one change: uracil, or U, substitutes for T.) Using these unique sets of four letters, arranged in three-letter words (the codons mentioned above), the cell is able to store and copy thousands of complex, specified files of genetic information within its hard drive (genome). We have already seen that the minimal quantity of specified complexity of theorized hypersimple bacteria seems to hover around 250 genes. (Recall that this is an extremely conservative estimate. The lower limit could actually be four times that or more.) Let's say that each gene has an average of 500 pairs of DNA letters. That multiplies out to 125,000 letters, arranged nicely in digitized biological files. But a recent study by Eugene Koonin suggests that the total of genetic letters (bases) in the simplest conceivable bacteria is likely to be in the minimal range of 318,000–562,000 letters.[4]

Let me take Koonin's lower limit and compare this mass of information with the text of the book you're reading. As I wrote each chapter, I used my word count function to see how long that chapter had grown. My self-imposed limit was around 6,000 words per chapter, which figures out to a bit over 30,000 letters on average. Using this as a guide, Koonin's typical genetic library for the simplest bacterium would have at least as many letters specified as ten of the longer chapters of this book combined—in other words, as long as this book, or longer! I have a hunch that most educated Americans have no idea that genomes of even the feeblest possible life forms must be packed with such amazing quantities of complex, specified DNA. Higher animals, with their 20,000 genes spooled away in the nucleus of the cell, would then total fifty times (or up to one hundred times) as much information—perhaps one hundred books or more.[5] This brute informational reality—the vast databases of informational stuff in DNA, RNA, and proteins—constitutes the second major engine of ID theory besides the irreducible complexity argument of Michael Behe. Such informational realities (whether in human books or DNA genomes) have been given a special name: CSI.

CSI and the Explanatory Filter

Please do not confuse *this* CSI with the popular television series by the same name. CSI in the nomenclature of Intelligent Design means "Complex Specified Information," a concept crucial to ID's system of design detection. Sometimes these informational patterns are simply called "specified complexity." The two terms are practically the same. This concept was developed mainly by William Dembski and Stephen Meyer in the period of 1992–1996.

How does one go about making the logical move from CSI (or specified complexity) to the intelligent cause that is responsible for its production? Dembski has shown one key way by linking CSI with another crucial idea, the "Explanatory Filter." The filter, which has gotten a lot of attention and criticism from ID's opponents, holds that any event or object in nature can be analyzed through a series of statistical or probabilistic tests, to see if it was likely caused *by a law of nature* or *by chance*. (Statistics and probability are Dembski's own area; he received one of his two Ph.D.s in mathematics.) Finally, failing both law and chance, the final test (or "specification test") is applied to see whether the event or object can be attributed to design. Even more than CSI, the Explanatory Filter has become an intense battle zone. More ink has been spilled, in print and metaphorically on the Internet, over Dembski and his filter than over any other topic in ID, with the exception of Michael Behe. The four major anti-ID bombs in 2004—the "bunker buster" books mentioned in chapter 4—were targeting Dembski's work as much as Behe's. All four books assaulted Dembski's arguments at length, but the world record goes to Mark Perakh, whose *Unintelligent Design* devotes his first full chapter, with an incredible *ninety-two pages*, to confronting ID's mathematician. (This is almost one-quarter of a 415-page book with fourteen chapters, tackling fourteen chosen targets.)

The temptation to anyone summarizing this huge debate is simply to quote some major criticisms from the anti-ID literature, then to paste in lengthy chunks of Dembski's key book, *The Design Revolution*, and call it quits. *The Design Revolution* is a supremely important book on the public debate over ID. It responds vigorously to over forty questions—many of them being hostile torpedoes aimed at blasting ID out of the water. I'll resist overquoting from *The Design Revolution*, but the reader is urged to obtain his own copy of the book and read it in tandem with this chapter to understand ID's robust response along this battlefront.[6]

This chapter will focus on the clashes over CSI and especially over the Explanatory Filter. To set the stage, I will quickly review the background story of the filter concept—how it was developed and how it works in its most modern form. Then we will walk through two of the stronger critiques of the filter and of Dembski's arguments generally.[7] With the attacks I will air Dembski's rebuttals, and in the conclusion of the chapter, we will witness in *The Design Revolution* some of his strongest criticisms to date against the Darwinian paradigm.

The Story of the Filter

Behind any great scientific idea or discovery lies a fascinating human story. This holds true for William Dembski and his Explanatory Filter. I spent most of chapter 9 in *Doubts about Darwin* telling the background story of Dembski—his academic training, which led to two Ph.D.s; his publication of the celebrated, peer-reviewed book *The Design Inference* through Cambridge University Press; the gist of his Explanatory Filter; and the roller-coaster ride of his career as a research professor at Baylor University.[8] Here I'll skip virtually all of that detail and zoom in on the Explanatory Filter by reliving a vivid conversation I enjoyed with Dembski after we crossed paths in the Seattle airport in August 1993. We were friends by this point, having met at a philosophy lecture at Princeton University in 1990 and having participated together in academic symposia since.[9] Our chat took place as I drove the two of us in a rental car to a meeting of the Ad Hoc Origins Committee—a forum for skeptics of Darwinian biology that was a forerunner of the Intelligent Design Movement. I doubt that he remembers the conversation, but it was unforgettable to me. After the usual chitchat, Bill casually mentioned to me the Explanatory Filter that had been germinating in his mind. I had no inkling of such a concept and asked him to explain it.

Of course I didn't take notes, since I was driving, but I've played back that conversation in my mind many times and feel I practically memorized it in essence. Dembski explained that if one wants to know, with logical and mathematical rigor, whether phenomenon X—some suspicious object or event—was produced by Intelligent Design, one can submit it to a battery of three filters. First, one asks if X has a high probability, such as would be produced very simply by the action of one or more laws of nature. If X is found to have a high probability (as in the case of a ball that is repeatedly dropped and always plunges toward the floor, with a probability near 100 percent), then it can be written off as fully explained by law or necessity (in the ball's case, gravity). However, if X remains unexplained, because it doesn't have a high probability, then X passes on to the next filter. One asks if it has a medium or moderately low probability, in which case one could easily explain X by reference to chance. Flipping a coin is one example; receiving a head can be explained by chance, since it has a medium probability of one out of two or 50 percent. (Later on, this level of chance was named "intermediate probability.")

The second filter, or the chance filter, can even snag moderately low probability events, and to illustrate this, I'll jump from the rental car to the poker nights (featuring a huge jar of pennies) that Ron and Janet enjoy with their friends Jason and Lori. It seems that Ron usually wins, so let's imagine that Ron is dealt a (literally) perfect hand—a royal flush in spades. The probability of this hand is 1 in 2,598,960, which is an awfully low probability, but it is not so low as to make it utterly implausible that someone would luck out with such a hand. In fact if there are 2,600,000 poker hands dealt in the U.S. this year (it may be many times that), it is very likely that someone, somewhere will be celebrating that elusive perfect hand, because there were so many trials. (The number of tries, in reaching a somewhat improbable event, is specifically figured into Dembski's reasoning. It's called a "probabilistic resource" of a "replicational" kind, and that's as far as I think I should get into the technical stuff.[10])

If we submit to the filter Ron's joyous triumph with a royal flush in spades, it passes through the first filter (it was not snagged by being a high probability or lawlike event), *but it was snagged by the second filter* by having a probability of merely one in 2.5×10^6—thus being a moderately low probability. In other words, Ron's hand could have been explained by chance. No one cheated, as far as we know; it was not by design.

Now let's imagine a scenario (admittedly extreme), a late night poker marathon, in which a total of twenty-five hands are dealt, and to the group's unending shock (and to increasingly justified suspicion) Ron is dealt a second royal flush in spades, then a third, then a fourth, and so on through the night in every single hand, in spite of the best efforts to shuffle the cards thoroughly between hands. Now, the probability of the entire imaginary event is vanishingly low—well below one in 10^{150}! How do we analyze this new hypothetical result by the filter?

It's time to return to the conversation in the car with Dembski, speeding along in Seattle. He explained that if phenomenon X is not caught by the first filter (high probability) or second filter (medium to moderately low probability), then by definition it is a "*very low* probability event" and is then referred to the third and final filter for the possibility that it was designed. But before we go to that third filter, we have to stop and ask: how low a probability is *very low*? In other words, at what very low probability level do you pass on from the second filter to the final filter? Personally, I would suspect "intelligent design" (crafty meddling) in a poker game if there were just two royal flushes in a row (with a probability of about 1 in 6×10^{12}), let alone twenty-five such consecutive hands. Yet given such a huge universe, with so much matter to jostle around and such a lengthy time span to work with, Dembski decided to be maximally conservative. He has set the standard for very low probability at an amazingly rare event: anything rarer than one in 10^{150}! This figure has a name: the "universal probability bound" of the Explanatory Filter. (A twentieth-century French mathematician named Emil Borel had settled for one in 10^{50} as his "universal probability bound"; others have suggested a bit higher than Borel.[11]) It would take about twenty-five consecutive perfect poker hands to reach Dembski's universal probability bound, but it would only take about eight consecutive perfect hands to reach Borel's limit. As I hinted above, Woodward's "universal probability bound in poker" is probably going to be two, or three hands at the most!

Technically speaking, any complex event or object tends to be exceedingly remote in its probability, so how do we keep from labeling as "designed" a merely random and meaningless outcome, such as the result of flipping a coin a thousand times? Such a pattern of coin flips would have a probability (one in 10^{300})[12] that goes well below even Dembski's extremely low probability bound, but obviously it is not intelligently guided. The answer is found in the third and last filter, which I held back in the explanation. As Bill wound up his three-filter explanation in the rental car, he said that if X is not a high probability event or a medium to moderately low probability event, it passes on to the third filter, which asks *if X conforms to some independently given ideal or specified pattern*. This is called the "specification filter."

Perhaps an example will bring this point into focus, but I need to switch from poker to "subtle messages in the kitchen." Letters spilling from a tipped-over box of Alpha Bits cereal on our breakfast counter will make interesting patterns, but it will not be expected to fall into the pattern: "TOM TAKE OUT THE GARBAGE." I will naturally interpret these thirty letters lined up as an unusually creative reminder from my wife and not ignore it as just an unusual, random accident. It is this *specified pattern*—whether in cereal messages, repeated royal flushes, or DNA sequences so vital for life—that the ultimate power of the filter is felt. Thus, if something is both "very improbable" (having passed through the first two filters) and is "specified" (having passed through the last filter), then we know, quite simply and directly, that some intelligence has designed it or shaped it. It did not happen by chance or sheer dumb luck.

To sum up, Dembski and Meyer, along with ID theorists generally, feel that science now has a principled and effective tool—the Explanatory Filter—with which to detect the involvement of intelligent causes in producing design in physical systems. When a cluster of gene sequences or a small group of protein (amino acid) sequences is fed through the Explanatory Filter, it is instantly confirmed to be the product of design and not law or chance (i.e., natural causes at work). In Dembski's book *No Free Lunch* (2002), he even took the bacterial flagellum and quantified its structure in such a way that it could be analyzed by the Explanatory Filter. The result will come as no surprise. Because of the flagellum's tight specification for function in its forty constituent proteins, along with its extremely low overall probability (calculated at 1 out of 10^{1170}, way beyond Dembski's probability bound), it was easily and overwhelmingly attributed to design by the filter. The probability of this flagellum having occurred by lucky selection of biological letters to form the forty proteins is equivalent to the probability of being dealt about 190 consecutive royal flushes![13]

The Explanatory Filter has been slightly modified over the past dozen or so years, but the basic idea remains the same; only the terminology is changed or adjusted. For example, the filter in its most current form, as published in *The Design Revolution*,[14] is unchanged in its essence. It simply employs now three "decision nodes":

1. ***Contingency***—if X is not contingent, that is, if X is not "iffy," then it is lawlike and is stopped there as fully explained. If it is contingent or iffy, it passes to the next node.
2. ***Complexity***—if X is not highly complex, in other words, if it has a greater probability of occurrence than 10^{150}, it is stopped in its tracks and attributed to chance. If the probability is so tiny and remote that it goes beyond this probability bound, it passes on to the next node.
3. ***Specification***—if X, now known to be highly complex and incredibly improbable, is found to be specified, conforming to an independently given pattern, *then it is attributed to design*. If not, then again chance can explain it.

So the newest form of the filter really is not that different from the rental car briefing I was fortunate to receive in 1993. Some Darwinian critics have made much of Dembski's changes in the filter, as if he cannot make up his mind or keeps fixing problems. The charge is false. Any good scientific idea will pass through a constant process of tweaking and improvement.

The rhetorical beauty of the filter is multifaceted: First, it is a conservative method, not jumping to the conclusion of design too quickly. Second, it also is rigorous—drawing the conclusion only after a methodical statistical analysis of probabilities. Third, it also signifies the religiously neutral thrust of Intelligent Design theory, in that it points to "an intelligence" generally and not to the specific identity of any agent or agents responsible. Fourth, it is a powerfully confirmed method to begin with, since it is rooted in the reality of empirical or inductive verification. Dembski points out that it never produces false results:

> The justification for this claim [of the reliability of the criterion] is a straightforward inductive generalization: in every instance where specified complexity is present and where the underlying causal story is known (i.e. where we are not just dealing with circumstantial evidence, but where, as it were, the video camera is running and any putative designer would be caught red-handed), it turns out design is present as well. . . . That's a bold and fundamental claim, so I'll restate it: *Where direct empirical corroboration is possible, design actually is present whenever specified complexity is present.*[15]

So if the negative empirical side of ID is seen in the work of Jonathan Wells, profoundly critical of the "proofs" of nature-driven macroevolution, then the positive empirical side is found to a great extent in CSI and the Explanatory Filter.

Darwinists Strike Back at Dembski and the Filter

It is to be expected that any new and controversial scientific idea, such as the Explanatory Filter, would be subjected to the harshest possible scrutiny and rigorous analysis by its opponents. That is what has happened, and such heavy criticism is certainly a necessary (if painful) part of the advance of ID concepts into the mainstream of science. On our related topic of CSI, it seems that an equivalent all-out attack has not been launched. There is some criticism of the concept of CSI (or specified complexity), which is used so heavily by design theorists, but the criticism is much more muted. This may reflect the fact that CSI functions as more of a basic, descriptive concept, used even by prominent non-ID scientists,[16] and it does not so evidently derive, on its own, the conclusion of design.

Strong criticism of new scientific concepts and methodologies is normal and helpful. What is less common (and less appropriate) is for opponents of a new scientific idea to attack the person developing the new idea. This is precisely what has happened in the case of William Dembski. Perhaps because of the perceived threat of his ideas, the strategy of attack among Darwinists against Dembski has gotten ugly and personal, with a sprinkling of sneers and contempt. Darwinian opponents often impute to Dembski (for example) an extreme self-confidence to the point of arrogance over the importance of his own ideas.[17] They repeatedly level a charge of Dembski using tedious (and excessive) "mathematical formalisms" or "mathematism as a tool of embellishment"[18] so as to overly impress lay readers with his scholarly erudition. I find these criticisms absurd, unfair, and entirely beside the point. They are, at best, indicative of the level of desperation of Darwinists in throwing everything they have at Dembski. But in any case, my purpose in the review of criticisms is to steer clear of personal mudslinging and concentrate on substantive objections.

The Problem of Mixed Causes

One of the most common complaints about the Explanatory Filter is that it wrongly focuses on one singular cause for analysis at a time, when in almost any case imaginable, more than one may be involved at the same time. Perakh describes it this way: "Dembski's categorical demarcation between law, chance, and design as the three independent causes does not seem to be realistic either, as it ignores multiple situations wherein either two or all three causes may be at play simultaneously."[19] Several other critics echo this point, especially Michael Ruse.[20] The gist is that when an event happens by mere chance (such as a single coin flip or Ron's single royal flush), it still entails the regularities of nature, such as the physical substance of the coins and the cards, the law of gravity, which regulates the coins or cards as they fly through the air, and so on. The same is true when an object is determined to be designed; it still partakes of law and chance to some extent. To clarify, let's listen in on Perakh:

> Consider Dembski's example of an archery competition. If an archer shot an arrow and hit a target, it is, according to Dembski, a specified event which definitely must be attributed to design. In Dembski's scheme, design excludes both chance and law. Can we really exclude law as a causal antecedent of the event in question? I submit that the archer's success was the result not of design alone, but of a combination of design and law. Indeed, the archer's skill manifests itself only in ensuring a certain velocity of the arrow at the moment it leaves the bow. This value of velocity is due to design. However, as soon as the arrow has separated from the bow, its further flight is governed by laws of mechanics. The specified event—the perfect hit—was due to both design and law. The arrow would not hit the target if any one of these two causal antecedents were absent. In this case design operates through law and would be impossible without law.[21]

At first blush, Perakh and his fellow critics seem to have spotted a fundamental flaw in the filter. But have they? When I first read Perakh's "mixed-cause" objection, it occurred to me that this was essentially a red herring. Recall the example I gave of a spilled cereal box, whose sugary letter bits on the counter spelled a reminder from my wife. Not one sane, scientifically literate person in the world would deny for a second that the bits of sugared cereal, shaped like English letters, were obeying the laws of physics and chemistry in their interior makeup and in their movements as they were spilled out and were then arranged in precise order by a loving hand. So the functioning of all normal physical laws was like a background given—an assumed substrate of the entire system being investigated. I did not observe the chemical structure of the sugars and carbohydrates in the bits of cereal and then blurt out, "That's it! It was the chemical laws of nutritional molecules working through chance that produced the message!" Such a conclusion would be absurd. Yes, the laws of physics and chemistry were at work with design in the cereal message, but the laws by themselves are not the crucial explanation that satisfies; their role is relatively trivial, and they do not suffice *alone* as a cogent explanation.

In the same way, a chemist and physicist can team up and study the precise physical workings of a neon sign. They can explain exhaustively just how the electrical circuitry works and why the neon glows in its reddish color range, obeying the laws of nature. But all their investigation will fail to explain the origin of the "Drink Coke" shapes that the neon tubes are twisted into. Physical law (or chance) will never adequately explain the written neon-light message.[22] This was the same basic point brought out by the Hungarian-British philosopher of science, Michael Polanyi, especially in his 1967 article, "Life Transcending Physics and Chemistry," as well as other publications, beginning as early as the late 1950s. Polanyi established a "method for establishing contingency via degrees of freedom in the 1960s. He employed this method to argue for the irreducibility of biology to physics and chemistry."[23] Just before his death, Polanyi had especially begun to probe the mystery of the information in DNA, and he was convinced that the knowledge of the physical laws governing the substance of DNA will not adequately explain the origin of the encoded message itself.[24]

Dembski himself disposes of the mixed-cause objection. In *The Design Revolution*, he first quotes Michael Ruse's objection to the filter. Ruse focuses on Ronald Fisher, one of the founders of the neo-Darwinian synthesis in the period before 1950. Ruse says that Fisher himself believed that "mutations come individually by chance, but that collectively they are governed by laws (and undoubtedly are governed by the laws of physics and chemistry in their production) and thus can provide the grist for selection (law) which produces order out of disorder (chance)." As an added bonus, Fisher "argued that everything was planned by his Anglican God!"[25] Ruse says at least two and maybe all three causes are hopelessly entangled. Dembski replies:

> Ruse is wrong that the Explanatory Filter separates necessity, chance and design into mutually exclusive and exhaustive categories. The filter models our ordinary practice of ascribing these modes of explanation. Of course all three can be run together. *But typically one of these modes of explanation predominates.* Is the rusted old automobile in your driveway designed? The rust and the automobile's beaten appearance are due to chance and necessity (weathering, gravity and a host of other unguided natural forces). But your automobile also exhibits design, which typically is the point of interest. What's more, by focusing on suitable aspects of the automobile, the filter detects that design. Ultimately, what enables the filter to detect design is specified complexity. The Explanatory Filter provides a user-friendly way to establish specified complexity. For that reason, the only way to refute the Explanatory Filter is to show that specified complexity is an inadequate criterion for detecting design.[26]

The Threat of False Positives

Dembski argues that the power and effectiveness of the Explanatory Filter is ultimately established by looking to see if it works. The filter is robust because it works—repeatedly, reliably, over and over when checked against the facts of nature. *In short, it has no known failure ever recorded.* That's a bold statement, but such a 100 percent success rate is pretty much a prerequisite for any theory that would become a reliable tool in the analytical work of science. I referred to this strength of the filter earlier, but Dembski recaptures the point:

> In thereby eliminating all material mechanisms, we are not saying that a phenomenon is inherently unexplainable. Rather, we are saying that material mechanisms don't explain it and that design does. *This conclusion of design derives not from an overactive imagination but simply from following the logic of induction where it leads: In cases where the underlying causal history is known, specified complexity does not occur without design.* Specified complexity, therefore, provides inductive support not merely for inexplicability in terms of material mechanisms but also for explicability in terms of design.[27]

In other words, the cause-effect patterns of the universe are unanimous. There is recorded no case, anywhere or at any time, when a fully known causal history of specified complexity was traced back to a natural explanation. This is important for several reasons. First, it confirms the legitimacy of the filter, since this procedure is new in its formulation and needs corroborating support to persuade the doubtful. Universal support from the known empirical case studies helps build strong support among scientists. Second, this inductive foundation of certainty about the filter (resting not on speculation but on the cause-effect structure of the universe) is an effective reply to the "God of the Gaps" (GoG) accusation. GoG arguments accuse design theorists of unnecessarily injecting God into a small, temporarily unresolved spot in an otherwise perfectly known tapestry of cause-effect relationships. The GoG bullets fly thick, for example, in Neil deGrasse Tyson's contemptuous dismissal of ID in *Natural History*.[28] In response to Tyson and other critics making GoG accusations, ID says, "No, this is no gap-filler enterprise. On the contrary, we are inferring the same kind of cause-effect relationship that is inductively observed, thousands of times a day, with no known empirical exception throughout the visible universe."

This is exactly where ID's critics seek to drop their biggest payload of explosives. If they can demonstrate one, or better yet, several, false positives that are produced by the Explanatory Filter, then the idea becomes a useless and irrelevant curiosity. If false positives stick, the filter is sunk. First, we need to ask: What is a false positive? Also, what is a false negative? Let's get the false negatives out of the way first.

False Negatives: No Problem

False negatives are no problem for the filter; both Darwinists and ID theorists agree that they are to be expected. A false negative is a false "no" to the key question: Was it designed? Intelligence can mimic natural processes, so intelligent action sometimes is not detected; it slips through the net. One example would be the scenario where Ron was dealt three consecutive royal flushes in a row. Any observer would strongly suspect intelligent meddling behind Ron's three spectacular hands. Let's assume, in line with my suspicion, that someone did meddle. It would be designed! But the ultraconservative Dembski filter would still not detect it, since the mathematical chance didn't reach the universal probability bound with its one chance in 10^{150}. Design happened, but the filter didn't declare "design" because of its very conservative nature.

Alternatively, imagine that when Ron got his first royal flush, the foursome stops to celebrate. While there is much high-fiving and guzzling of raspberry iced tea, Jason sneaks a peek at the next (shuffled) deck. To his horror and suspicion, he sees that Ron is bound to get another royal flush on the second hand! Suspecting monkey business, Jason slyly rearranges the cards to produce an unimpressive, random-looking outcome for Ron. So the next hand—quite ordinary looking—was designed by Jason, but no one suspects it, and no application of the filter could ever detect it. In both poker examples, the key thing to grasp is that this false negative phenomenon is no threat to the filter; it is totally expected.

False Positives: Case #1

False positives are another world entirely. The idea of a false positive emerging from the filter is a false "yes" to whether something was indeed designed. If X passes all the way through the filter and is found to be of incredible complexity (hopelessly improbable) yet highly specified, then the label "designed" will be slapped onto X. But what if X turns out *not to be designed* in matter of fact? Several of Dembski's critics announce such false positives and say the filter is shipwrecked. Dembski says that each of these examples is *not a false positive*, and thus the Darwinists' arguments collapse. Who is right?

We will single out two supposed false positives that have been put forward as cases that disprove the filter. The first one is called the "Fibonacci series"—a special number series by which some plants space their leaves on a branch. Darwinists such as Gert Korthof argue that the daily output of Fibonacci numbers, in the spacing of leaves of certain species, represents a designed event. It is equivalent to receiving a string of prime numbers from outer space. Only a computer can mimic such a Fibonacci output, using a mathematical formula. Since we see this complex and specified event happening over and over under the botanist's nose, with no one intelligently intervening, it is as if the Fibonacci series emerged from Dembski's filter, labeled "designed," only to realize that it all happened without intelligence.

Dembski's response is simple. To say (as Korthof) that the arrangement of leaves in a Fibonacci pattern happened by a "perfectly natural process," is to equivocate regarding the word *natural.* The key question is: What is the *event of interest* that is being detected as designed? Where is intelligence involved? Is it in the day-to-day functioning of the Fibonacci-spaced leaf system, programmed in a plant's biochemical interior? Or is the event of interest here "the structuring event that arranged biological systems so that they could output Fibonacci sequences in the first place"? Even generously granting that perhaps the software in the cell that could output Fibonacci patterns of leaves might be fairly simple and might arise naturally, the software only works inside the vast and intricately complex hardware system of the preexisting plant cell. Dembski adds that the "simplest functioning cell is staggeringly complex, exhibiting layer upon layer of specified complexity and therefore design."[29] Thus, the natural operation of a thing is confused with its own designed origin. This confusion, says Dembski, is rampant in the literature that criticizes ID.

False Positives: Case #2

The other major false positive that is mentioned in several books (especially those involving Niall Shanks) is the phenomenon known as Bénard cells, which are a honeycomb pattern of hexagonal cells of moving water, produced when a wafer-thin film of water is encased between two glass plates and heat is applied to the underside. The typical honeycomb pattern of water motion is caused spontaneously, and yet the cells can vary somewhat; there is a high degree of flexibility and variability—hence real complexity, beyond the Dembski universal probability bound. Such cells of circulating flow are even said to be detectable on the surface of the sun. The key threat to Dembski's filter, according to Shanks, is that Bénard cells, "forming in accord with dumb, natural mechanisms, manifest complex specified information." Later on he repeats the claim that Bénard cells "manifest CSI and they arise from natural unintelligent causes."[30] Since any entity with a true designation as CSI presumably would be both (1) *complex*, having passed through the first two filters, and (2) *specified*, having successfully passed through the third and final filter, the implication is that Bénard cells indeed have been awarded the label "designed" as they exited the triple filter. Since these cells arise naturally in their flattened watery environment between glass plates, and not by intelligence, it is claimed that the filter has misled us.

First, a personal reaction, then a comment from a published interaction. When I read the chapter in Shanks's book, I thought, "Weak argument—what possible relevance or parallel do such cells of swirling water have with the digitally specified informational sequences in DNA or proteins?" It brought to mind what I had heard years ago, when early ID researchers began to formulate the information argument. Some defenders of prebiotic evolution pointed to the soapy swirl patterns that spontaneously formed when pulling the plug in a bathtub. The bad news for origin-of-life researchers is that such swirling patterns indeed are ordered, but order has little to do with information. Indeed, order or orderliness in some senses is almost the *opposite of information* because order implies simple master structures, or regularity, or periodicity, like a salt crystal. Information in the cell, on the other hand, is virtually the opposite; it is profoundly *aperiodic*—it does not contain any simple repeating patterns. So when I read about the Bénard cells, I thought, "Soapy swirls in the bathtub strike back, only now without soap and sandwiched between plates of glass!"

These cells seem bereft of true marks of complex information, and furthermore they are certainly not "independently specified" as if for some target function. Thus they cannot possibly be classified as CSI. If anything, the filter snags them at the first or second level: they are explicable as simple patterns, driven by lawlike processes (the setup, done right, always produces the same basic pattern), although manifesting slight variances, due to the variables (chance is the factor here) of the external environment. They appear not to be proceeding to the final filter at all. However, if Shanks insists in pushing matters down to the molecular level—in going to the trillions of water molecules bobbing about in their unique swirls, so that he can show that the collective movements of these molecules are complex enough to go way beyond the Dembski bound in the improbability of their pattern, then I say, "Fine. Let's go to that level. And as soon as we arrive there, the lovely Bénard cells get quickly escorted off to the side at that level, into the analytical bin labeled *chance*, due to their flunking the specification test." In other words, they cannot possibly pass through the final filter and be legitimately called CSI.

Cornelius Hunter, a biophysicist reviewing *Why Intelligent Design Fails*, includes a comment about Shanks's Bénard cells: "Much of the criticism, however, does not seem fatal to ID. Niall Shanks and Istan Karsai argue that complexity can arise from purely local mechanisms. But . . . Bénard cells . . . require a clever apparatus."[31] In other words, with the constrained laboratory structure provided by two glass sheets, perfectly water-filled thin space between, and carefully modulated heat source, evenly applied, what does this artificial environment have to do with the rough-and-tumble of natural environments at work, shaping matter as it can? Is it not possible that the limited (noninformational) order that appears in the swirls is a predictable result of highly constrained, intelligent (i.e., unnatural) structured conditions supplied by the experimenters? I agree with Hunter that these examples do not touch the relevant issues of CSI and the Explanatory Filter. Both alleged "false positives" fade away upon close inspection. The filter is again vindicated.

Concluding with a Point

As we exit the smoke-filled battleground surrounding William Dembski, it is wise to pan out to the broader landscape of what we've learned about the universe from the discussion of CSI and the filter. One of his most powerful chapters in *The Design Revolution* (and there are many)[32] is called "Information Ex Nihilo." He places an epigraph at the beginning of that chapter: "Is nature complete in the sense of possessing all the capacities needed to bring about the information-rich structures that we see in the world and especially in biology? Or are there informational aspects of the world that nature alone cannot bridge but require the guidance of an intelligence?" In answering the questions, Dembski goes back to the fiery hypercompressed dot of quarks that existed just after the big bang and asks if "all the possibilities for complex living forms like us were in some sense present at that earlier moment in time." Many assume the answer to be "yes," but the early history of the universe "still doesn't tell us how we got here or whether nature had sufficient creative power to produce us apart from design."[33] A leading philosopher, Holmes Rolston, the environmental philosopher at Colorado State University and recipient of the Templeton Prize, says in *Genes, Genesis and God* that there's no sense in which human beings or any other sort of creature are latent in single-celled organisms. Says Rolston, to assert that life is somehow lurking in chemical substances or that complex life-forms are lurking in simple biological systems is "an act of faith."[34]

Yet from a "barren, searing, and tempestuous cauldron" of the early Earth, life emerged, says Dembski. But how did that happen? What caused it? He says, "Now we can conjecture that blind natural forces all by themselves made it happen. But if so, how can we know it? And if not, how can we know that? According to the theory of intelligent design, the specified complexity exhibited in living forms convincingly demonstrates that blind natural forces could not by themselves have produced those forms but that their emergence also required the contribution of a designing intelligence."[35] At this point, Dembski is ready to introduce a new twist on an old theme—ex nihilo creation. He is not looking at the universe itself, coming from nothing, but rather he is plumbing the source of information—to see if it too comes from nothing:

> The design found in nature therefore demonstrates that nature is incomplete. In other words, nature exhibits design that nature is unable to account for. What's more, since the design in nature is identified through specified complexity, and since specified complexity is a form of information and since this form of information is beyond the capacity of nature, it follows that specified complexity and the design it signifies is information ex nihilo. That is, it's information that cannot be derived from natural forces acting on preexisting stuff. Indeed, to attribute the specified complexity in biological systems to natural forces is like saying that Scrabble pieces have the power to arrange themselves into meaningful sentences. The absurdity is equally palpable in both cases. Only in evolutionary biology the absurdity has been repeated so often that we no longer recognize it.

The "ex nihilo" in Dembski's conception does not mean "from nothing" in an absolute sense but rather "from nothing in nature itself." In the rest of this chapter, he argues (and I agree) that the proper contrast, contra Niall Shanks, is not between natural causes and "supernatural miraculous" causes but between natural causes and intelligent causes. When we act, as humans, doing things nature can never do, we don't break the laws of nature; we simply act as intelligent agents, making choices, producing small mountains of specified complexity each day. So natural law is not broken in the process. Nevertheless, even though Intelligent Design implies no contradiction of natural laws, "it demonstrates a fundamental limitation of natural laws, namely that they are incomplete."

The upshot of Dembski's discussion is his provocative conclusion: "Intelligent design regards intelligence as an irreducible feature of reality. Consequently, it regards any attempt to subsume intelligent agency under natural causes as fundamentally misguided and regards the natural laws that characterize natural causes as fundamentally incomplete."

In a sense, scientific research into CSI and the Explanatory Filter is today's supremely dangerous science—dangerous to long-held traditions about how to investigate our world at all levels. This is so because it threatens to recast and reorganize the long-negotiated hierarchy between natural science and the study of other (or broader) realities, including scientific method, philosophy, and even theology, as we shall see.

11

Unexpected Allies

Cosmologists and Atheologians

Unexpected allies! This theme brings to mind an incorrigible, unpredictable, practically indescribable intellectual—an unrelenting scourge of Darwinism—who burst onto the scene quite unexpectedly in the summer of 1996. His abrupt debut came with the publication of "The Deniable Darwin" in *Commentary*.[1] I'm referring to the nonreligious Jewish intellectual David Berlinski, who received his Ph.D. in philosophy from Princeton and has achieved international renown as a historian of mathematics and physics. This cosmopolitan thinker and gifted wordsmith does not exactly fit the religious profile of Barbara Forrest's would-be ID theocrats. His religion, he once quipped, amounts to "having a good time all the time." His *Commentary* essay, which provoked a deluge of indignant letters to the editor, including howls of pain from several of the world's most prominent Darwinists, argued that macroevolutionary theory was so lacking in mathematical rigor that it hardly deserved to be called a scientific theory.

Two months later, because of the enormous response, a special issue carried the double explosion of letters (positive and negative) and allowed Berlinski to explain in detail why these responses tended to confirm his thesis rather than blunt it. In the decade since then, regarding the research program of ID, he has written sometimes as a collaborator with design theorists in their quest for greater mathematical precision. On other occasions he has written essays in which he donned the mantle of friendly critic, poking and prodding in potential weak areas. In each case, this unexpected ally has added a distinctive French tang to ID discourse (he lives in Paris) and has helped enormously in the development of the ID paradigm.[2]

A second kind of odd ally helped in my own writing and publishing of *Doubts about Darwin*. As I wrote it originally as a Ph.D. dissertation and later circulated copies to academic reviewers, I assumed that my history of ID, retold from a sympathetic stance, would inevitably trigger scowls and strong criticism from scholars of an atheist or agnostic bent. That proved wrong many times. Six key academics who do not believe in God helped in the research, review, endorsement, and publication of my book. (Two of these, sociologist Steve Fuller and MIT professor emeritus Murray Eden, gave enthusiastic blurbs.[3]) These six nontheist colleagues in academia, including both natural scientists and social scientists, were indeed my unexpected allies—crucial *heroes of the plot* in the story of my book seeing the light of day.

I must also point out a secret trio of unexpected allies: the three confidential peer-reviewers of Stephen Meyer's article in the *Proceedings of the Biological Society of Washington* (August 2004). According to the journal editor, Richard Sternberg, these credentialed biologists gave numerous suggestions for change or clarification in the manuscript. Crucially, while not agreeing with the essence of Meyer's position, all three felt that Meyer's article was very well constructed and articulated, presented a legitimate point of view, and thus the essay deserved to be published, since this area is badly in need of fresh thinking. The fury that arose from the Smithsonian after this article was published is an enduring symbol of the courage of these peer-reviewers (still confidential to this day) who, as allies of open dialogue, gave the green light for publication.

Strange Allies Emerge

These three opening examples remind us that scientists and other university professors have also become central heroes of the plot, and very unexpected allies, in the debate over Intelligent Design. Both sides of this struggle have been reinforced in their arguments by contributions of such researchers. From the standpoint of those who see Intelligent Design as "spiffed up creationism," any such scientific support for the ID side may seem not just unexpected but also frustrating and even galling. Nevertheless, reports of such scientific endorsements and help for ID work weaken the position that ID is "religion, not science."

On the other side, it can be surprising if not shocking to see Darwinists receive extensive religious support of various kinds from (1) theologians, (2) church leaders, (3) scholars who are evangelical Christians, and (4) atheologians. The fourth type of religious supporter, the atheologians, are scholars who brandish *theological arguments against ID*, while portraying them as scientific or logical arguments. The atheological supporters are the *other* unexpected allies we will survey in this chapter. (We will meet some theological and evangelical supporters of Darwin in chapter 12.) This spectacle of religious or theological support for evolution is not new. Prominent churchmen in Darwin's day jumped to his defense. Even conservative Presbyterian scholars such as James McCosh and B. B. Warfield played an important role in quelling the early opposition to evolution and easing its acceptance at Princeton University and Princeton Theological Seminary in the late 1800s. That trend lasts to today, with the involvement of Langdon Gilkey and other theologians as key supporters of evolution (and opponents of creationism) in the 1980s and beyond.[4]

On the other side is the spectacle of support or cooperation flowing to the Design Movement from a wide spectrum of scientists, mathematicians, and philosophers—even including evolutionary biologists and paleontologists, as we shall see. Many non-ID researchers have helped in strengthening design arguments and vindicating their scientific accuracy.[5] In *Doubts about Darwin* I told how two key scientists did more than merely show respect to Phillip Johnson as he was researching Darwinism in the period 1987–1991. They came to his assistance through discussion, reading of book drafts, and the offering of suggestions for improvement. They appreciated his carefully thought-out point of view, while declining to embrace his thesis. These two who helped Johnson are the late British Museum curator Colin Patterson and the influential paleontologist David Raup.[6]

It would be misleading to imply that Raup and Patterson were the only fact-checkers and dialogue partners with whom Phillip Johnson worked while he was writing his critique of Darwinism. Dozens of other scientists outside of Johnson's circle played a key part in the shaping of his work. In fact, over the past two decades Johnson was careful to send drafts of his work to hundreds of scientific peer-reviewers outside of the ID network, including Gould himself. This diligent and humble intellectual habit of seeking feedback from the strongest possible critics fully vindicates Johnson's intellectual integrity against the repeated charge that he was an "outsider sticking his unscientific nose into matters he knows nothing about." Even the Berkeley science professors who read Johnson's first book draft on Darwinism and then came out to discuss it in his faculty colloquium in September 1988 played a significant role in getting the earliest ID discussions going (see chapter 4 of

Doubts about Darwin for this revealing encounter). In this sense, Johnson's unexpected professorial allies undoubtedly number in the hundreds![7]

In the twenty-first century, civility among scientists toward ID has receded somewhat. As the debate heated, leading Darwinists drew the line in the sand, insisted on a "no concession" policy, and substituted sneers and distortions for intellectual interaction.[8] So the fairly congenial spirit seen in the late 1980s and 1990s is now in partial eclipse, as some scientists who formerly engaged in relaxed discussion of ID now have pulled back and fallen mum. But that uptightness may be passing, and there are some crucial exceptions to this pattern.

Physics and Cosmology

One cannot read up on the Intelligent Design Movement without soon coming across unexpected allies in the form of physicists and cosmologists who provide evidences and arguments for design from their respective areas. In *Doubts about Darwin* I focused almost exclusively on the *inner space* of biology and only slightly mentioned the *outer space* dimension of ID theory. I had good reasons for doing this.[9] As we are well into the twenty-first century, it has become clear that ID discussions are connecting regularly with the study of the very fundamental structures of physical reality and the history of the universe, conducted mainly by astronomers and physicists. ID opponents see the importance of this topic, and their recent critiques have featured cosmology to some extent. Mark Perakh's *Unintelligent Design* includes a chapter on Hugh Ross, one on Fred Heeren, and one on Israeli physicist Gerald Schroeder. Taken together, about seventy-five pages (one-third of his space devoted to science topics) cover these three who use primarily cosmological or physical evidences to point to an intelligent designer. This pattern is reflected in Niall Shanks's *God, the Devil, and Darwin* as well as *Why Intelligent Design Fails*, edited by Matt Young and Taner Edis. In both of these books, one chapter is devoted to debunking the case for design in cosmology.[10] Clearly, this is a relevant part of the controversy! This discussion twirls like an elegant binary star system: the two shining topics that dance around each other are the big bang and the universe's fine-tuning.

The Big Bang—Friend of ID?

The development of the big bang as a hypothesis for the origin of the universe can be traced to its earliest theoretical hints in the early 1900s, but it was decades later when the Russian-American scientist George Gamow (pronounced "Gam-off") chose the term *big bang*. (An opponent, Fred Hoyle, used the term in 1946 as a phrase of disparagement.) So history records that Gamow, assisted by a few other scientists, wrote the first theoretical descriptions and predictions of the big bang in 1948.

Though glimmers of this theory came from Einstein's equations of general relativity in 1917, the first experimental evidence came in 1914 when Vesto Slipher announced that he had detected the movement of about a dozen galaxies flying away from us. This pattern of fleeing galaxies was confirmed by other astronomers, and especially by Edwin Hubble, whose discovery in the late 1920s of the expansion of the universe is now viewed as one of the modern turning points in astronomy. These revolutionary new ideas were captured in the early 1930s by Belgian mathematician and Catholic priest George Lemaitre, an acquaintance of Einstein. Lemaitre's concept of "fireworks exploding from a primordial atom" was the earliest theory of the big bang before it was named such.[11] In the 1950s and 1960s there was a competing theory of origins, the Steady State Theory, which held out hope for an eternally old universe. But then came a key discovery in the mid-1960s, when Arno Penzias and Robert Wilson, working with an early radio telescope, detected the first signs of the *cosmic microwave background radiation*. This radiation that pervades the universe was interpreted as *fossil radiation* of the big bang—the afterglow from the initial cosmic expansion. After the Penzias-Wilson discovery, the scientific community rapidly accepted the big bang, but this was not without grumbling from certain quarters.

Controversy, Controversy!

The big bang played a strange role in regard to ID, almost as a forerunner. Twenty years before the birth of ID in the critical fires of Michael Denton, the idea of evidence in nature pointing beyond herself was already bubbling to the surface in the theory Gamow had christened. By the time ID was being formulated in the late 1980s, the big bang was cited as highly relevant to design theory in two ways. First, ID theorists observed that the big bang was a theory with religious implications, but these implications did not prevent the theory from receiving a fair hearing. *What mattered was whether it was supported by good evidence, not whether it had religious implications*. Likewise, ID theory may have religious implications, but that also should not disqualify it automatically. Second, the big bang is one of the first modern scientific theories that provided important scientific hints of true design of the universe. The big bang scenario posits a basic *creation event* that naturally raises the question of its own cause. What or who triggered the bang? At a minimum, it seems to *suggest* a designer as one possible explanation, even though it does not necessarily *demand* a personal designer.

Controversy has never fully left the big bang theory. The idea of a continuous expansion of the cosmos, and the implicit starting point to which it pointed, was originally opposed by Einstein himself, though by the 1930s he had reconciled himself to the idea. The big bang seemed to be confirmed by various measurements of the cosmic microwave background radiation (CMB) in the past fifteen years. Current measurements imply that the universe is neither eternal nor infinite; rather, it burst into existence about 13.8 billion years ago. In recent years, the theory has been declared by some researchers as overwhelmingly vindicated by every conceivable measurement or observation. The vast majority of writing and speaking on the matter by astronomers and others proclaim the success of the big bang with a supremely confident tone.[12]

Nevertheless, a small but vocal minority of scientists feel that the big bang is untenable. One recent history of the theory by Simon Singh mentions the healthy role of these critics, many of whom hold to a "Quasi-Steady State model." Singh adds, "Cosmologists who continue to support this minority view are fiercely proud of their role in challenging the big bang orthodoxy. Indeed, Fred Hoyle, who died in 2001, went to his grave in the firm belief that the Quasi-Steady State model was correct and that the big bang model was wrong."[13] Publishing an open letter in the *New Scientist* (May 22, 2004), thirty-four scientists—many of them nontheists—signed a statement claiming that the theory is hobbled by three newer theoretical factors: dark matter, dark energy, and Alan Guth's "cosmic inflation" hypothesis.[14] In spite of such points of dissent, probably well over 90 percent of ID scientists accept the big bang scenario. Yet, what matters here is not anyone's belief but rather the pro-ID implications flowing from the theory.

The Strange Case of Robert Jastrow

One physicist renowned for diving into the implications of this theory is the agnostic space scientist Robert Jastrow. His comments on the big bang in his classic *God and the Astronomers* form the launching pad for our discussion. Jastrow details "the enormity of the problem" that the big bang presented to science: "Science has proven that the Universe exploded into being at a certain moment. Was the Universe created out of nothing, or was it gathered together out of pre-existing materials?" Now comes Jastrow's key to the problem's *enormity*: "And science cannot answer these questions, because, according to the astronomers, in the first moments of its existence the Universe was compressed to an extraordinary degree, and consumed by the heat of a fire beyond human imagination." Jastrow explains that the "shock of that instant must have destroyed every particle of evidence that could have yielded a clue to the cause of the great explosion." In other words, science is at least *somewhat stymied* on this point, and this is an "exceedingly strange development, unexpected by all but the theologians."[15]

Theistic (or deistic) implications of the big bang are well-known, and they have been pointed out in print by dozens of scientists, including Michael Behe.[16] Jastrow is famous for boldly addressing the possibility of an intelligent origin of the cosmos. He asks whether the big bang has led science to the doorstep of a designer, closing his book with these now-famous words: "*For the scientist who has lived by his faith in the power of reason, the story ends like a bad dream.* He has scaled the mountains of ignorance; he is about to conquer the highest peak; as he pulls himself over the final rock, he is greeted by a band of theologians who have been sitting there for centuries."[17]

One might guess that Jastrow throws up his hands at this point and says, "Okay, okay! The universe was created—it came from a preexisting intelligence!" But that's not quite right. Less quoted but equally revealing of Jastrow's mind-set is the book's opening. There he says: "When a scientist writes about God, his colleagues assume he is either over the hill or going bonkers. In my case it should be understood from the start that I am an agnostic in religious matters. My views on the question are close to those of Darwin who wrote, 'My theology is a simple muddle. I cannot look at the universe as the result of pure chance, yet I see no evidence of beneficent design in the details.'"[18]

So by his own description, Jastrow is sitting on the theoretical fence. He states that he has not embraced theistic belief, but he seems pulled in two directions. On one side, he is tugged toward the *suggested possibility* of a higher mind behind the universe—perhaps the same being that the "theologians at the summit" have referred to for centuries.[19] On the other side, he feels pulled by his commitment to a naturalistic approach to investigating the physics of the universe, based on his sense of how science works. Jastrow's unwavering commitment to methodological naturalism continued right on through the recording of a short video interview in 2003, which is in the bonus section of the 2004 documentary *The Privileged Planet*. His tension surfaces in a poignant way when he acknowledges that his commitment to a naturalistic framework of thought seems to conflict with hints of design that are apparent from the evidence.

The oddity here is that the scientific community has shown a kindly tolerance of Jastrow's speculation about the big bang as containing hints of theism. Why such a gentle treatment of Jastrow, especially since ID theorists who point to DNA and cellular nanomachines as empirical markers of design are treated as dangerous nonscientific ideologues? The key seems to be that Jastrow's possible (suspected?) designer operates *outside* the current space-time continuum. The hypothetical deity is pictured as only launching the universe—as merely setting it up. There is no suggestion that he ever intrudes into the day-to-day workings of the universe after it starts to expand, to shape matter into DNA or cellular complexity.

A Fine-Tuned Universe?

The implication of a designer from the big bang is one pole of our cosmology debate. The other pole is the set of physical constants and quantities that make up what is commonly called the *fine-tuning* of the universe. Sometimes the name *anthropic principle* is brought in as a quasi-synonym for fine-tuning. *Anthropic* comes from the Greek word *anthropos*, which means "man," and thus the universe is said to have qualities that make it a congenial habitat for carbon-based life-forms, specifically *humans*. Without these fine-tuned conditions, humankind could not physically exist. It is misleading to think that anthropic coincidences somehow prove by themselves the role of a designer outside (preexisting) the cosmos. The original idea of the anthropic principle as developed by John Barrow and Frank Tipler in their groundbreaking work, *The Anthropic Cosmological Principle*, was not used as an argument for an intelligent designer. In fact, philosopher William Lane Craig argues that this massive book on fine-tuning actually sought to end the teleological argument for design.[20]

To tackle fine-tuning, let's start with the basics. When you arrive at the fine-tuned universe (hereafter referred to as FTU), you exit the theory-scented realm of the big bang (a "highly plausible hypothesis" that is accepted by most scientists, but not all) and you step into a new realm of *startling scientific facts*. Here, no one denies the astonishing data that have accumulated. This young field can be traced to a presentation by Brandon Carter in 1974, and each year since then we seem to be hearing more examples, more specific cases, of the FTU. Many scientists and science writers have published lists of factors or constants that are fine-tuned; it is not uncommon to find lists with over one hundred factors. *So the facts are there—in plain sight*. The only controversy is how to interpret the FTU discovery. First, let's review some calibrated constants and quantities of the cosmos; then we will proceed into the war zone of interpretation.

Scientists who write about fine-tuning typically begin with two sets of constants and settings in nature: the precise mass of the subatomic particles and the calibrated strengths of the four forces of nature (gravity, electromagnetism, the strong force, and the weak force). For example, if these forces were tweaked just the tiniest fraction of a percent stronger or weaker, then unpleasant results would tumble out of the hypothetical equations, among them being that stars couldn't burn steadily or for great duration, heavier elements would not be fused into existence, and the flourishing of higher biological life would become virtually impossible. From such lofty and technical fine-tuning aspects—of matter, energy, and the forces of nature—one can descend down into the humble and humdrum lowlands of such local fine-tuning issues as the atmosphere of Earth or even the orbital placement of our planet. We are smack-dab in the middle of what astronomers call the "Goldilocks zone"—a mathematical space around the sun that is *not too hot* (too close to the sun as Venus), *not too cold* (too far away from the sun as Mars), but *just right*. In *The Privileged Planet* (both in the video but far more in the book), many other parameters of the universe and the Earth system are explained by Guillermo Gonzalez and Jay Richards: right type of star as the sun, position in the galactic habitable zone, ideal plate tectonics, abundance of water, position and size of the moon, and much more.[21]

One super-dramatic example of fine-tuning is the carefully balanced rate of the cosmic expansion over an estimated 13.8 billion years. This rate of expansion had to be simultaneously *not too fast* (too-fast expansion would prevent galaxies from forming) and *not too slow* (too-slow expansion would cause the universe to collapse upon itself in a fiery death spasm shortly after the bang). Is there a large middle ground, a safe speed zone, between the two cliffs of *too fast* or *too slow*? Amazingly, there is almost no middle ground at all. Instead of a narrow butte with two cliffs on each side, it is a long knife edge. How carefully is the rate of expansion tuned? One Berkeley astrophysicist stated that the expansion is fine-tuned to sixty decimal places, adding that this level of fine-tuning is "crazy."[22] Some scientists have said that the expansion rate, in view of the recently added factor of *dark energy*, seems to have put the level of fine-tuning for the expansion rate of the universe as drastically more precisely fine-tuned than previously estimated: one part in 10^{120}! This is a level of precision that more than boggles the mind. It is a trillion trillion trillion trillion trillion times "crazier than crazy."[23]

There exist a wide array of books surveying the fine-tuning discovery and the various interpretations, both those of an intelligent designer (including Neoplatonic or deistic designers) and those of a more materialistic or even pantheistic nature, one could consult literally dozens of excellent books. The reader might start with John Leslie's enduring classic, *Universes*.[24]

The Meanings of Fine-Tuning

Fine-tuning seems to be a many-splendored thing, a phenomenon given to wildly divergent interpretations. Scientists affiliated with (or supportive of) ID point to the growing list of fine-tuned factors as having far too remote a chance to have clicked into place by luck. Since there is no known law that determines these factors, the only alternative is deliberate choice of those qualities and quantities—true design. Opposing this conclusion we find many scientists who adhere to naturalism. They tend to explain away such factors and quantities in the FTU as fortunate but inconsequential. Some say that the FTU is either "physically inevitable" (cosmic qualities and quantities could not be otherwise—that's just the way the universe had to be). A good example of this point of view is captured in one paragraph in the online Wikipedia discussion of the anthropic principle:

> Stephen Hawking suggest[s] that our universe is much less "special" than the proponents of the anthropic principle claim it is. According to Hawking, there is a 98% chance that a universe of a type as ours will come from a Big Bang. Further, using the basic wave function of the universe as basis, Hawking's equations indicate that such a universe can come into existence without relation to anything prior to it, meaning that it could come out of nothing. As of 2004, however, these publications and the theories in them are still subject to scientific debate, and in the past, Hawking himself has asked, "What is it that breathes fire into the equations and makes a universe for them to describe? . . . Why does the universe go to all the bother of existing?" (Hawking 1988).[25]

The second major way of explaining the FTU is that perhaps we are in the one superlatively tuned universe that blasted into existence among many others. In other words, perhaps our cosmos turned out to be incredibly beautifully constructed, by chance, *among many zillions of parallel universes that exist apart from ours*. This theorized population of universes is sometimes called the Multiverse. As early as 1989, this idea was already a preferred way to dodge the religious implications of the FTU when John Leslie wrote *Universes*. By 1997, when Timothy Ferris's *The Whole Shebang* appeared,[26] more and more researchers were embracing this view. Those who have not followed this area of science are often shocked to hear that many scientists seriously propose that our entire vast universe is just one among myriads (perhaps countless trillions) of other universes, bobbing around in some hyperdimensional realm, a sort of *hyperspace-hyperuniverse*. The visual imagery connected to this concept is often rather striking and can even entail mother-daughter connections between different universes. Picture, for example, a baby universe popping out of (or bubbling up from) a more mature universe.

If there are trillions of universes in this hyperspace manifold, then according to the naturalistic scenario, eventually one of these universes will come into existence with all of its necessary factors and constants just right to permit carbon-based life to flourish. That universe would be like the holder of a winning lottery ticket in some mega-lottery, while the other universes (at least a vast majority) would be "loser" universes. Some had such a clumsy set of physical factors they collapsed and died shortly after their birth. Others might have factors that permit them to expand and endure but that never produced anything more interesting than hydrogen gas.

Is there any evidence for the existence of such other universes? Scientists have admitted that observers probably would be unable, *even in principle*, to detect another universe. We certainly have no evidence at this time for their existence. Nevertheless, they are extremely popular as an explanatory acid to dissolve the troubling designer inference. ID theorists (and other critics of the Multiverse) have pointed out that even to get a "baby universe" to pop out of a mother universe and start expanding, such a rare hypothesized event would itself require an incredible set of fine-tuned factors to happen. So (to coin a new acronym) even the cosmic delivery room would demand an FTBP—Fine-Tuned Birthing Process, which itself is evidence of a designer! One could say with justification that such a runaway population of universes is accepted into one's belief system simply by faith. The Multiverse thus is commended only as an ad hoc add-on hypothesis to help deal with the mystery of the FTU. It seems that the scientific principle of parsimony, known as Ockham's razor, would act to cancel or slice away this *bloated ontology*,[27] this runaway ensemble of unobservable parallel universes. Naturalists reply, "On the contrary, ID's inferred nonbodily designer is the one that violates Ockham's razor. It—the hypothesized designer of high complexity—needs to be sliced away by the razor!" So we seem to have arrived at another argumentative stalemate. Or have we?

The Gonzalez-Richards Thesis

This stalemate, I argue, has recently been broken by a book discussed earlier: *The Privileged Planet*. This work by Guillermo Gonzalez and Jay Richards appeared both in print and in a one-hour video documentary that summarized the highlights of the book. Both the book and the video shed important light on this topic.

The Privileged Planet broke new ground, questioning the Copernican Principle and the Principle of Mediocrity, both of which implied that there is nothing special about our earth or its place in the cosmos. This new line of research focused on fine-tuning as a pattern of evidence not only linked to the ability of the universe to support carbon-based life but, equally and more importantly, as a pattern that supported and enabled scientific measurement and discovery to take place here on the planet where intelligent life exists. In short, the universe and especially our planet are not just fine-tuned for life; they are also fine-tuned for science.

Let me elaborate. Most people who have gazed at nature over the centuries have noted two positive qualities, both of them related to nature's pleasant aspects: (1) the *simple beauty and visual grandeur* of nature (for example, a peacock's feathers or Florida's sunset hues of orange and aqua spattered gloriously over the Gulf of Mexico), or (2) the multidimensional *comfortable nature of our globe as a fit habitat* for us humans and for our fellow animal and plant species. Yet virtually no one, before the thesis of Gonzalez and Richards exploded onto the scene, noticed (3) the wondrous functional aspect of nature in permitting us ideal conditions to glimpse, to discover, to measure, and to scrutinize nature herself. In short, the *design of nature for science itself* has lain generally unnoticed and undiscussed until very recently.

At one point in *The Privileged Planet* video, this new dimension becomes visually striking. A 3-D special effects sequence takes the viewer soaring into the star-clogged, dusty, hostile interior of the arms of our galaxy, where supernovas and nebulas would not just pose a danger but would also greatly limit viewability. Not a great place for the solar system if one wants to do astronomy! Then, in the next sequence the same viewer platform whizzes into the relatively open and spacious cavern of lightly populated space between the two arms of our galaxy, where the Earth and its solar system is actually located. There, we literally see ourselves suspended in an ideal location—both for safety and for prime viewing of both the universe and our own galaxy. For astronomy, the positioning could not be better!

Many other factors drew the attention of the coauthors of *The Privileged Planet*, but it is time to arise from particulars to the general point. If fine-tuning was just a chance event (either in one universe or in this universe among billions of other flawed, boring universes), it might explain the fine-tuning that led to the emergence of intelligent life. But law or chance would not explain the extra, gratuitous, unnecessary-for-survival fine-tuning for scientific discovery to take place. That newer, much more shocking dimension of fine-tuning suggests an intellect, a master designer who is interested in more than just the flourishing of life. This designer appears to have planned and ordered nature and the Earth so that scientific discovery would be enabled.

Here, it seems, the *chance* or *law-determined* explanations of fine-tuning begin to struggle mightily. Not that there are no objections that the philosophical naturalist can lodge. In fact, Gonzalez and Richards respond to fourteen such objections in their next-to-last chapter. The reader is invited to review these objections and the replies given by the authors to see how thoroughly the objections were dealt with. To use a baseball analogy, *The Privileged Planet* is the ace's scientific curveball—a wicked sinking slider thrown past the opponents of ID. In June of 2004, Gonzalez and Richards were shocked to open *Nature*, the world's most prestigious science journal, and find a respectful and somewhat positive review of their book! Even

Simon Conway-Morris of Cambridge University, the highly respected pioneer of the Burgess Shale Cambrian fossils discussed in chapter 7, weighed in on behalf of this audacious thesis: "In a book of magnificent sweep and daring Guillermo Gonzalez and Jay Richards drive home the arguments that the old cliché of no place like home is eerily true of Earth. Not only that, but if the scientific method was to emerge anywhere, the Earth is about as suitable as you can get. Gonzalez and Richards have flung down the gauntlet. Let the debate begin; it is a question that involves us all."[28]

Atheologians to the Rescue

If ID theorists have been helped unexpectedly by findings from physics and cosmology, then the other major surprise of this chapter is that the Darwinian cause has enjoyed a steady flow of support from theologians and *atheologians*. I describe the latter group as scholars lodging *theological* points against design *as if they were knockdown scientific arguments*. I shall focus on the atheologians in this section; we will listen to theologians in the next chapter.

The Strange Case of Richard Dawkins

Richard Dawkins is the world's most staunch and widely published figure who is both defending Darwinism and attacking traditional theism. His early fame was established by the mid-1980s through *The Selfish Gene* and *The Blind Watchmaker*. The latter book helped launch ID when Phillip Johnson read it simultaneously with Denton's critique of Darwinism. During the 1990s, after Johnson published *Darwin on Trial* and later Behe began to make waves, Dawkins said virtually nothing about Intelligent Design. However, that changed after 2000, when he wrote several short pieces against ID, in which he joined the chorus of fantasy-spinners. For example, in a feisty foreword written for Niall Shanks, Dawkins says, "Intelligent design 'theory' is pernicious nonsense which needs to be neutralized before irreparable damage is done to American education." He is in Shanks's amen corner when it comes to the more extreme ID fear-fantasies: "For rather odd historical reasons, evolution has become a battlefield on which the forces of enlightenment confront the dark powers of ignorance and regression." Wow—dark powers! More important and substantive than Dawkins's rhetorical framing of ID theorists as villains is his repetition of his favorite "knockdown argument" against design. This argument is not new. In *The Blind Watchmaker*, Dawkins set up the same theological argument against any notion of creation. Here it is again, in Shanks's foreword:

> Darwinism and design are both, on the face of it, candidate explanations for specified complexity. But design is fatally wounded by infinite regress. Darwinism comes through unscathed. Designers must be statistically improbable like their creations, and they therefore cannot provide an ultimate explanation. Specified complexity is the phenomenon we seek to explain. It is obviously futile to try to explain it simply by specifying even greater complexity. Darwinism really does explain it in terms of something simpler—which in turn is explained in terms of something simpler still and so on back to primeval simplicity. Design may be the temporarily correct explanation for some particular manifestation of specified complexity such as a car or a washing machine. But it can never be the ultimate explanation. Only Darwinian natural selection . . . is even a candidate as an ultimate explanation.[29]

What are we to make of this seemingly simple, rule-based argument? He seems to take an earlier argument in *The Blind Watchmaker*—basically, "Who made the designer?"—and simply reframes it, announcing that "the statistical improbability of a designer" is a fatal problem. But surely Dawkins, a very bright man and a marvelous writer, should realize that he veers here into blatant circular argumentation. He simply assumes—without any evidence-based argument or philosophical proof—that no intelligence can ever exist who is a necessary (uncaused) being. This amounts to a religious or metaphysical claim, equivalent to asserting a naturalistic worldview to be the only rational approach to scientific investigation. In other words, to say that an *undesigned designer is simply impossible* is to settle the scientific question of ID on the flimsiest of foundations: a biased and untested philosophical rule. Dawkins does not make an empirical case against ID, although in writing this foreword, he endorses what Shanks has written. Most significant is that his own anti-ID lunge amounts to a bare assertion. The archbishop has spoken: science, by its very nature, must be the search for ever-simpler causes of events in the universe! Design theorists reply two ways:

1. Is there any good reason to assume, *a priori*, that a highly intelligent and powerful being might not exist apart from the space-time continuum? The problem of an eternal regress, asking "who made that designer, and then who made that designer's designer," simply dissolves the moment we recognize the *possibility* of an *undesigned* or *uncaused* designer. If the designer be (for instance) God, and if God is defined as an uncaused being who transcends time and space, then to say that such a being is "impossible" is simply to shout loudly, "ID is bunk because I don't believe in God!" One can say this, expressing a point of view, but no one with scientific and logical finesse would dare call that a scientific argument. Thus Dawkins's gambit is simply

to announce that his view of reality shall now serve to define the task of science and shall limit the explanatory options available. How sensitive to the leading of empirical evidence is that? Or, most crucially, where is the empirical evidence—or airtight logic—to ground such a bold declaration? Obviously, there is none.

2. Design theorists would challenge the notion of a lesser-evolving-to-greater pattern of cause and effect as the "rule in science." This is simply an ad hoc declaration and without consistent empirical support. Dawkins admits that examples exist of *greater producing lesser*—and that this is a normal pattern of the universe's causality. But in the end, he seems to put maximum weight of confidence on the power of natural selection. Yet that is exactly the most empirically empty part of the Darwinian scenario! *ID argues that, in fact, there is no good empirical evidence of lesser producing greater in any significant degree.* In the final analysis, there is amazingly little weight at all in Dawkins's knockdown argument; the argument is lighter than a helium balloon.

Dumb-Design Accusations

A second and much more common argument against ID from atheologians is to point out problems in nature that declare that no intelligent designer has created the complexities of nature or continues to control the affairs of nature. Often this takes the form of listing examples of evil and waste in nature. Because of limited space, I will pass over the evil-and-waste argument.[30] More famous in the context of ID are alleged cases of dumb design in biology, which a genius designer (God) would never have tolerated. Some flaw in X (e.g., in a molecular system or an organ) is listed; then it is asserted that any tidy-minded designer would never have produced X. Also called the "suboptimal design argument," it amounts to a theological claim that "God wouldn't have done it that way."

This type of argument is nothing new. It is one of the great thrusts in Darwin's own rhetorical project, as he sought to topple the design hypothesis that dominated biology in his day. He was ceaseless in pointing out the cruelties, wastes, and other oddities and imperfections of nature. Even before he published *The Origin of the Species*, this was a popular intellectual trend among scientists and laypeople alike. ID theorist and biophysicist Cornelius Hunter has written extensively in *Darwin's God* on the heavy usage of such theological argument. Hunter argues that there was an amazing "God-focus" in Darwinists' writings ever since Darwin.[31]

The most famous example of dumb design is the *backward retinas* of vertebrate animals. All classes of vertebrates have retinal cells (rods and cones) wired in backward so that the tips face the back of the eyeball and the nerve fibers arise into the eye's fluids and descend en masse through the famous *blind spot*. The blind spot, we are told, is a stupid design feature that threatens creatures' survival. Worse still, the backward retinal cells don't meet the design druthers of certain Darwinists. Why not have them face the light, as they do in the eyes of octopi? The conclusion of *poor design* is thus said to prove *no design*; this clumsy arrangement is just what dumb natural selection would leave behind. Therefore, the retinal cells, the eyeball, the optic nerve, and all the rest of the vision system (including brain centers that decode the signals) must not have been designed!

Such pickiness strikes me as a superficially clever argument, but the more focused one's thought becomes, the weaker the argument seems on every level. First, why is such a design (cells wired facing backward) assumed to be second-rate? Biologists in recent years have suggested excellent reasons for seeing this structure as vastly superior.[32] Second, "poor design" accusations don't eliminate the design inference, they just criticize it as inferior. My first car, a 1970 fastback Toyota Corolla, was a wonderful car, surely designed by intelligent engineers, but I could make a list of things I didn't like. Yet even ten hours of nitpicking would not lead to the inference it was not designed. What is almost always overlooked (or deliberately unmentioned) is that any event of design deals with a wide range of practical constraints. The Corolla could have been built to last one hundred years, but that would have added considerably to its cost. Its power and speed could have been increased, but fuel economy would then be sacrificed. As Dembski asks, "Is there an even minimally sensible reason for insisting that design theorists must demonstrate optimal design in nature? Critics of intelligent design (e.g., the late Stephen Jay Gould) often suggest that any purported cosmic designer would only design optimally. But that is a theological rather than a scientific claim."[33]

Thus the dumb-design accusation (or insistence on *optimal design*) is based on notions about God and not on any flaw in the design inference. Any specific design act in the real world (including retinas) entails a complex decision that makes trade-offs of one constraint against another. William Dembski shreds this type of argument in his book *The Design Revolution*:

> Now my point here is not that the human eye can't be improved or is in some ultimate sense optimal. My point, rather, is that simply drawing attention to the inverted retina is not a reason to think that eyes with that structure are suboptimal. Indeed, there are no concrete proposals on the table for how the human eye might be improved that can also guarantee no loss in speed, sensitivity and resolution. There's also an irony here worth noting;

> the very visual system that is supposed to be so poorly designed and that no self-respecting designer would have constructed is nonetheless good enough to tell us that the eye is inferior. We study the eye by means of the eye. And yet the information that the eye gives us is supposed to show that the eye is inferior. This is one of many cases in evolutionary biology in which scientists bite the hand that feeds them.[34]

The dismissal of design based on poor design reaches a fantastic extreme in a caustic essay by astronomer Neil deGrasse Tyson in the November 2005 issue of *Natural History*. In my critique I shall not demean the notable achievements of Tyson as an astronomer, astrophysicist, and extraordinarily gifted science writer. His voluminous writing has led readers into a vivid and gripping exploration of the secrets of the universe. Astrophysics has been opened up to the educated nonscientist with rare clarity and power. But in his essay "The Perimeter of Ignorance," Tyson sadly descends into an extreme version—resembling a parody—of simplistic atheology.

Tyson opens his essay with a high-speed review of the achievements of Newton, Laplace, Galileo, and others, with emphasis on our progressive liberation from the grip of religion upon science. He then takes a swipe at the idea of a designer through a verbal rhapsody on natural evil. We face fearsome dangers both terrestrial (e.g., bears, snowdrifts, and viruses) and celestial (e.g., "explosions of supermassive stars"). He says the "evidence points to the fact that we occupy not a well-mannered clockwork universe, but a destructive, violent, and hostile zoo."[35] When Tyson shortly thereafter adds that "the universe wants to kill us all," his implication seems clear, if unprinted: *only an idiot would look upon such a scene and say it was designed by an intelligent (good) designer*. With all the flourish and finesse of a village atheist, Tyson dismisses thousands of years of profound wisdom of philosophers and theologians who have written on the causes and implications of natural evil. In addition, the knowledgeable reader will inevitably ask Tyson: what about the dozens of protective factors about the solar system that have wonderfully shielded us from such dangers, such as our system's placement between two galactic arms, and the role of the large outer planets and our large moon, acting as vacuum sweepers to catch encroaching comets and meteors? Dozens of such positive factors are ignored; evil is accentuated; balanced intellectual discussion is banished. The function and tone of Tyson's argument is similar to that of a crafty trial lawyer trying to sway a jury rather than a careful, reflective investigator of nature.

Having hinted against design using natural evil, Tyson then takes the gloves off, but he shoots himself in the process with his own out-of-control Uzzi. In my nearly twenty years of reviewing attacks on ID theory (starting with published reviews of Denton's *Evolution: A Theory in Crisis*), I have never experienced what happened while reading the last two sections of Tyson's essay. In the midst of reading, I literally started laughing at his arguments. I kept shaking my head at his poor-design accusations passed off as scientific argumentation. I wondered, "Is this a *Saturday Night Live* spoof on dumb-design arguments?" Clearly it wasn't. Let me share Tyson's partial list of what is wrong with human design; I must remind you that I'm not making these up:

1. We are not water creatures, thus subjecting us to possible drowning.
2. We have useless appendages like the pinky toenail and an appendix.
3. Our knees and spines are poorly designed.
4. Unlike cars, we don't have built-in "biogauges" (equivalent to lights on the dashboard) to warn us of dangers like high blood pressure.
5. Our eyes are punky detectors for astrophysics—no ultraviolet or infrared—or radar detection "to spot police radar detectors"!
6. Unlike birds, who navigate through the magnetite built into their heads, we don't have such gifted noggins, and thus we need maps to get around an unfamiliar city.
7. We don't have gills [see point 1], and we should have more than two arms to be more productive.

Tyson's exercise in absurdity suddenly shifts to a passionate declaration, "Stupid design . . . may not be nature's default, but it's ubiquitous. Yet people seem to enjoy thinking that our bodies, our minds, and even our universe represent pinnacles of form and reason. Maybe it's a good antidepressant to think so. But it's not science—not now, not in the past, not ever."[36]

Tyson's list is presented as a fleshing out of what he calls "clunky, goofy, impractical, or unworkable" things that "reflect the absence of intelligence,"[37] but this list manifests in dazzling fashion the absence of intelligent rhetoric. As an attempt at serious persuasion, it backfires. The Atheologians' Club is probably enjoying Tyson's essay as a bit of rousing entertainment. However, those looking for serious reasons to reject Intelligent Design may experience the opposite effect. A philosopher I know who teaches at an elite private college and is not a member of the ID Movement reflected on the strangeness of Tyson's essay. He asked, "What is going on in science when this kind of discourse can be published as serious argumentation in a national science magazine?"

Even though atheologians such as Tyson certainly bend their rhetorical efforts to defend Darwinism by cataloging poor design, the intellectual flaws of this diversionary argument are now evident. Will these scholars wind up unintentionally fueling the advance of Intelligent Design? I suspect they are already doing that.

Now that we have traversed the exotic terrain of unexpected allies in cosmology and atheology, we must rise once again to the plains of wide-ranging inquiry and place our core question before us once again: has the war of rhetoric against ID managed to halt the renegades in their tracks, or has it sped up the unraveling of the Darwinian paradigm? That is the final and central question that we must confront in our last chapter.

12

Are We at the Tipping Point?

Theses, Flashbacks, and Questions

When I was serving as a photo intelligence officer in the U.S. Air Force during the early 1970s, one of my jobs was to oversee the writing of BDA reports—electronic summaries of Bomb Damage Assessment—after inspecting aerial photos of targets. Now that we have finished our overview of the vast scientific struggle over the origin of nature's amazing complexity, the moment has arrived when a BDA summary is called for. We must assess where things stand after a decade of Darwinism striking back at Intelligent Design—with growing intensity in the late 1990s and then escalating to a ferocious pitch after 2004.

We have seen wave upon wave of rhetorical bombs and rockets rain down upon ID concepts and arguments, but the design theorists were not passive or silent. Their own rockets of countercriticisms repeatedly struck the weakest points in Darwinian theory, with emphasis on two key issues: (1) Is there any empirical evidence that the mechanism of mutation-selection, touted as the "generator of complexity," can create new body plans and new files of genetic information? (2) Should naturalistic philosophy, which banishes "intelligent causes" from scientists' explanatory toolboxes, be viewed as a necessary starting point for good science? ID's leaders viewed these two components as the primary buttresses that held the entire Darwinian paradigm together. As soon as it was shown that these ideas possessed mainly ideological support and little or no evidential backing, then the reigning paradigm would be fatally weakened and its collapse would be inevitable. Thus, in light of this bigger framework, this chapter will go beyond a final BDA summary to assess the "State of the Paradigm." The ensuing discussion will culminate in a report comprised of three theses. Next I will go on to share some recent experiences that have built my confidence in the projections, and finally I will test the theses by means of key questions that will surface final objections.

Looking Back, Looking to the Future

During our journey, we spied out a number of battlefields. We first reviewed the volley of ID critiques in the period beginning in 1996 (such as *Darwin's Black Box* and *Icons of Evolution*), and we highlighted the later ID-inspired crash courses (such as the peer-reviewed books *Debating Design* from Cambridge University Press). We also spotted the strategic importance of the video shown on PBS stations, *Unlocking the Mystery of Life*. We have seen and sampled the evolutionists' harsh counterattacks upon Michael Behe, Jonathan Wells, William Dembski, Stephen Meyer, and other encroaching scientists. In a pair of chapters we toured the labyrinth of origin-of-life research that has proved so frustrating to scientific materialists. We surveyed the skirmishes over the fossil record and the Cambrian explosion, and we reviewed the evidence that is suggestive of design within cosmology and physics.

In the course of our battle reports, we met a number of key defenders of Darwinism, including Kenneth Miller, Robert Pennock, and the three authors of the Oxford Press attacks on ID.[1] We could feel the ground shake as four bunker-buster bombs blasted the ID strongholds and dug their rhetorical craters in 2004. We were astonished at the Darwinists' panicky fantasy-themes, which accused ID's scientists of paving the way for theocracy. We sensed the vehemence and passion as bullets of atheologians rained down on design theory, painting the universe as an evil, maniacal killer and holding up to view the alleged "clunky features" of our bodies' "dumb design."

We have also noted the equal intensity and conviction of ID's returned fire, with design theorists taking advantage of each attack by exposing fresh evidentiary weaknesses in Darwinism and showing that it is not gaps in evidence or a "God of the Gaps" argument that underlies the design inference. Rather, it is the very opposite: design is inferred in light of the continual growth and accumulation of new evidence. ID responders singled out

Darwinists' routine use of logical fallacies, and even *theological* arguments, in their attacks on design. Behe especially led the charge in exposing evolutionary "Just So Stories" that passed for scientifically rigorous answers to his own irreducible complexity arguments.

Clearly, significant damage has been done on both sides of the battle lines. That makes assessment especially tricky. To frame the analysis, let us recall the image of a towering fortress, used in chapter 3 to picture the *publicly assumed plausibility of nature-driven macroevolution*. This stronghold, protected by impressive walls and powerful armaments, was the target of ID's campaign of persuasion. In terms of that metaphor, a prime question is whether any major Darwinian explosive charges may have become neutralized or may even have backfired, causing friendly fire damage. We need to see whether any outer walls of that stronghold have been breached. And ultimately, we must descend to the bedrock question: are there any clear indicators that some sort of paradigm shift—at the level of macroevolution—is underway, either (1) away from neo-Darwinism to some other naturalistic theory (such as that hinted at by Gerd Muller and Stuart Newman in *Origination of Organismal Form*), or (2) from neo-Darwinism to a blended paradigm of Intelligent Design combined with Darwinian microevolution?

Alternatively, one can turn my key questions around and rephrase them with an aggressive and "triumphant over ID" spirit, as Kenneth Miller might: *There never was a threat to the fortress of Darwinian truth. And why should anyone pay serious attention to ID now that their arguments have been brought out in the open and demolished? How long can ID cover up its sense of disarray, now that stunning fossil evidence for macroevolution has come to light, and now that we have golden examples of irreducibly complex systems evolving right in our Petri dishes? ID may not have been completely crushed, but the failure of design theory to win the support of significant numbers of evolutionary biologists, or even biologists in general, shows that there is little scientific merit in all this fuss. The danger has passed.*

The Tipping Point and the Theses

By phrasing the rhetoric so strongly here from Miller's side, I seek to capture the drastically adversarial spirit that has come to dominate the scene. Yet despite this hostility of verbiage that is gushed forth daily, I am convinced that we have arrived at the historical tipping point, where Darwinism has entered a time of accelerating hemorrhaging of the educated public's assumption that it remains a plausible theory. In coming years, more than at any time in its previous history of a century and a half, college and university graduates across the world will begin to doubt that the fantastic claims of Darwinism about the complexity-building powers of nature are really resting on sound evidence and careful testing. This sense of where history is and where it is heading will be stated in terms of a three-point thesis. It is rather risky to foreshadow the future, but this triple thesis seeks to sketch the path that inexorably seems to be opening up:

1. The paradigm crisis has arrived. The Kuhnian crisis in Darwinism so prominently symbolized in Michael Denton's *Evolution: A Theory in Crisis* is no longer on the horizon but rather has come to powerfully dominate the current scene. It is now brute historical reality, virtually undeniable. Even though the PR machine of macroevolution proclaims, "All is well; the invaders' arguments are in tatters," this fantasy-theme is woefully out of touch with scientific reality. On the contrary, key walls of the aging paradigm have in fact been breached, and more are being breached by new data as I write this chapter (see below).

2. Neo-Darwinism has bought time but will soon be replaced by two or more competitors. Here I venture into BDA summary and paradigm prediction. The fierce Darwinian assault on ID has slowed down the acceptance of several key design arguments, has achieved some significant media support (even among conservative columnists George Will and Charles Krauthammer),[2] and has gained a few immediate political windfalls (such as the Dover ruling in December 2005). However, seriously flawed rhetorical practices of Darwinists have started backfiring. I have in mind especially the prevalence of dumb-design and other theological arguments, the widespread use of "Just So Stories," and the use of *ad hominem* attacks, the genetic fallacy, and "poisoning the well" in assaults on ID.[3] Sometimes the "backfiring rhetoric" was made especially embarrassing, as it was noted by non-ID scholars such as philosophers Del Ratzsch and Neil Manson (see chapter 4). Ratzsch performed a lengthy surgical analysis of Niall Shanks's book *God, the Devil and Darwin* and exposed an astonishing array of errors, distortions, vilifications and faulty reasoning. This is one of many cases when frantic anti-ID rhetoric exploded in the Darwinist lap. See the appendix for an excerpt from Ratzsch's review, "How Not to Critque ID Theory." These rhetorical habits have served to highlight the weaknesses in the current paradigm. Just as important, the cell's irreducible complexity—ID's prime argument—is proving horrendously stubborn to any nature-driven explanation. It is highly recalcitrant—not yielding an inch to Darwinian explanatory efforts, in spite of Kenneth Miller's vigorous PR efforts to convey the opposite impression.[4] Franklin Harold's book, *The Way of the Cell,* expresses the typical lament over this unyielding problem: "We should reject, as a matter of principle, the substitution of intelligent design for the dialogue of chance and necessity; but we must concede that there are presently no detailed Darwinian accounts of the evolution of any biochemical system, only a variety of wishful speculations."[5]

In view of these trends in rhetoric and research, combined with the accelerating international spread of ID, its penetration of peer-reviewed journals, and the ebbing plausibility of nature-driven macroevolution, the current decline of neo-Darwinism will continue steadily and may well accelerate into a catastrophic implosion within ten years. Eventually (by 2025 at the latest) the aging paradigm will yield its dominance to at least two competitors. The first competitor or, more likely, *competitors* will be newer (but much less crisply defined) naturalistic paradigms—perhaps one will include ideas of Stuart Kauffman, but most likely one will be built upon the work of newer evolutionary developmental biology (evo/devo) scholars such as Muller and Newman and other *Origination of Organismal Form* contributors. The second type of competitor will be a blended paradigm in which ID is fully integrated with Darwinian microevolution. (Some in this ID paradigm—like Behe—will work in agreement with the hypothesis of common ancestry, and others will continue to do research to demonstrate the empirical case against common ancestry.) Both the new-naturalistic and ID-oriented paradigms will vie for scientific allegiance in a lively and healthy competition, and this multiparadigm mode of research may last for a number of years.

3. Reasons for the inevitable decline of Darwinism. The current paradigm in biology, as I suggested above, will not survive much longer. In my admittedly biased point of view, *it simply cannot survive since it is fatally hobbled with two very implausible claims.* First, it ascribes to natural selection of random mutations (and to other unintelligent mechanisms) a massive spectacle in nature—the coordinated writing of 20,000 digital files on the DNA hard drives of higher species. Second, it points to selection as the sculptor of new bodies and other organic structures of plants or animals. Some biologists may say that this is just a single problem (production of new morphology *through* new genetic material). But recent critiques from researchers of diverse backgrounds show that the one problem has now split into two separate mysteries. Let me explain this "one problem split into two," beginning with the side of new DNA.

As we saw earlier, just a primitive early cell would require more than *300,000* bits of information coded in correct order; we said this was equivalent to all the words and letters in this book. Compare that quantity with a typical current arthropod, the fruit fly *Drosophila melanogaster*, which has *180 million* base pairs in its DNA, which is tightly spooled and tucked away in its nucleus.[6] This is equivalent (in information content) to about 600 books the length of this one. If the rise of a single book of information (the minimum for a barely functioning cell) is deemed extremely implausible through unintelligent processes (see chapters 8 and 9), then how much more is that true of the whole libraries of books possessed by arthropods and other phyla bursting forth in the Cambrian? The origin of dense quantities of coded information arising from mutations served up by unintelligent nature seems a task far beyond staggering, *unless there is good evidence that even one gene has ever been so crafted by the power of selection*. Sadly for neo-Darwinism, there is very little experimental evidence of nature crafting even one genuinely new gene, let alone the thousands of genes that would be needed to support the complex phyla that swarm in the Cambrian. How can my previous sentence, seemingly a radical denial of current orthodoxy, be sustained scientifically? Does not this go against everything we have been told in our biology classes?

Amazingly, from ongoing experiments with bacteria, it seems that the power of selection is limited when two or more mutations are both needed in order for the organism to benefit. Even when bacteria are given thousands of generations to evolve, they are unable to do so when evolution requires two steps. *Need three mutations? The data say: no way*. The "needing-two-mutations-at-the-same-time-stops-evolution" finding was brought to light at the August 2005 Uncommon Dissent Forum in a riveting talk by biologist Ralph Seelke, whose own experimental data on bacteria supports this conclusion.[7] Seelke's proposed limit on mutations in a single system—"two at the most when both are required"—has practical implications. There is some evidence that unique single-gene sequences can be altered by one or a few mutations to arrive at a new function of that gene (see the literature on the evolution of the enzyme called *nylonase*). But to write fresh gene sequences time after time *to support a multiprotein system* seems well beyond the reach of chance. If the rise of (for instance) a five-gene system, which codes for five coordinated proteins, appears nearly hopeless given the restriction to unintelligent causes, what about the 20,000 coordinated genes in the mammalian genomes? Thanks to the theoretical work of Stephen Meyer, ID maintains great respect for the *uniform cause* convention in science (the present is the key to the past), and points out we only have experience of such quantities of coordinated information arising from intelligence. The alternative idea—the chance/necessity explanation for the rise of genomes of 20,000 files—does not rest on compelling evidence in the real world. It rests primarily on a general confidence derived from a naturalistic view of nature.

Now for the second side of the "one problem split into two." Even if, contrary to the facts above, the origin of new DNA was explained, that still does not mean that the arrival of new structures or new body forms has been comfortably explained. This insight can be upsetting to defenders of the reigning paradigm, which assumes that natural selection of new genes simply birthed new morphology. Yet increasingly, the rise of new DNA sequences and the rise of new body structure are treated as separate issues. For example, the scholars who contributed to *Origination of Organismal Form* (hereafter *OOF*) have clearly detached the origin of new DNA sequences from the origination of body plans. In our new era, influenced increasingly by this *OOF* perspective, more and more biologists of all stripes are rejecting the central claim that natural selection acting on random genetic variation has the power to generate new body form.

As a result of this second problem of morphology (on top of natural selection's "origin of DNA problem"), the current paradigm appears to face a double whammy. As Muller and Newman pointed out in *OOF*, natural selection of genetic mutations can certainly accomplish the variation of existing form, which is commonly called *microevolution*. This is where Darwinism seems to work reasonably well; Darwinian science here is rock solid. But the classic gene-centered paradigm, say the editors of *OOF*, "has no theory of the generative";[8] that is, it fails to shed light on the macroevolution of new body structures.

Some may say that biology without Darwinian *macroevolution* is like English class without verbs or stellar physics without gravity. Often the words of Dobzhansky mentioned in chapter 4—"Nothing in biology makes sense except in the light of evolution"—are invoked. However, in connection with Dobzhansky-type claims, Philip Skell, an emeritus professor at Penn State University and member of the National Academy of Science, raised eyebrows with his opinion piece in *The Scientist* (August 29, 2005). Skell stated his thesis in two key paragraphs:

> Darwin's theory of evolution offers a sweeping explanation of the history of life, from the earliest microscopic organisms billions of years ago to all the plants and animals around us today. Much of the evidence that might have established the theory on an unshakable empirical foundation, however, remains lost in the distant past. For instance, Darwin hoped we would discover transitional precursors to the animal forms that appear abruptly in the Cambrian strata. Since then we have found many ancient fossils—even exquisitely preserved soft-bodied creatures—but none are credible ancestors to the Cambrian animals.
>
> Despite this and other difficulties, the modern form of Darwin's theory has been raised to its present high status because it's said to be the cornerstone of modern experimental biology. But is that correct? "While the great majority of biologists would probably agree with Theodosius Dobzhansky's dictum that 'nothing in biology makes sense except in the light of evolution,' most can conduct their work quite happily without particular reference to evolutionary ideas," A. S. Wilkins, editor of the journal *BioEssays*, wrote in 2000. "Evolution would appear to be the indispensable unifying idea and, at the same time, a highly superfluous one."

Skell reports on his informal survey of seventy eminent researchers who explore the biology of the living world. He asked "if they would have done their work differently if they had thought Darwin's theory was wrong. The responses were all the same: No." He adds, "From my conversations with leading researchers it had become clear that modern experimental biology gains its strength from the availability of new instruments and methodologies, not from an immersion in historical biology." (For a bit more on the Skell controversy and his interaction with critical letters, see the sidebar on page 180.)

Philip Skell and *The Scientist*

A number of scientists wrote letters, arguing that Skell had overlooked aspects of the usefulness of evolutionary theory. Philip Skell was permitted to respond to his critics:

> My essay about Darwinism and modern experimental biology has stirred up a lively discussion, but the responses still provide no evidence that evolutionary theory is the cornerstone of experimental biology. Comparative physiology and comparative genomics have certainly been fruitful, but comparative biology originated before Darwin and owes nothing to his theory. Before the publication of *The Origin of Species* in 1859, comparative biology focused mainly on morphology because physiology and biochemistry were in their infancy and genomics lay in the future; but the extension of a comparative approach to these subdisciplines depended on the development of new methodologies and instruments, not on evolutionary theory and immersion in historical biology.
>
> One letter mentions directed molecular evolution as a technique to discover antibodies, enzymes, and drugs. Like comparative biology, this has certainly been fruitful, but it is not an application of Darwinian evolution—it is the modern molecular equivalent of classical breeding. Long before Darwin, breeders used artificial selection to develop improved strains of crops and livestock. Darwin extrapolated this in an attempt to explain the origin of new species, but he did not invent the process of artificial selection itself.
>
> It is noteworthy that not one of these critics has detailed an example where Darwin's Grand Paradigm Theory guided researchers to their goals. In fact, most innovations are not guided by grand paradigms but by far more modest, testable hypotheses. Recognizing this, neither medical schools nor pharmaceutical firms maintain divisions of evolutionary science. The fabulous advances in experimental biology over the past century have had a core dependence on the development of new methodologies and instruments, not on intensive immersion in historical biology

> and Darwin's theory, which attempted to historicize the meager documentation.
>
> Evolution is not an observable characteristic of living organisms. What modern experimental biologists study are the mechanisms by which living organisms maintain their stability, without evolving. Organisms oscillate about a median state; and if they deviate significantly from that state, they die. It has been research on these mechanisms of stability, not research guided by Darwin's theory, which has produced the major fruits of modern biology and medicine. And so I ask again: Why do we invoke Darwin?[9]

That concludes my three theses. My growing confidence in the theses is rooted not only in the evidence sketched here, nor even in two years of research into the attacks and counterattacks along the front lines of this debate. What has clinched my confidence is a rich kaleidoscope of experiences that I had during several events that took place in the months just before this book went to press. One event was the first European ID conference on October 22–23, 2005, held in Prague, Czech Republic, and the other event was an ID televised "debate." Each event convinced me, in different ways, that ID is far closer to its rhetorical goal than I realized and that the rational defense of Darwinism has become far weaker than I thought. Let me recapture those moments.

Epiphanies Signaling Paradigm Flux

Not long ago, I found myself in a televised discussion of ID with a college president whom I'll call "Dr. Smith." I had read Dr. Smith's very harsh public attack on ID, which had been published just weeks before the broadcast. Echoing what he had written, Smith said that his college would consider teaching about ID in philosophy or religion classes but never in science classes because "ID is faith, not science."

Going on the offense, I explained the evidence for design from the astounding complexity of the flagellar motor and other complex systems and brought out a few other scientific points in order to quash Dr. Smith's thesis that ID is "faith, not science." The epiphany, however, took place not on air but off the air—between two segments. I leaned over to the college president and expressed my interest in knowing about the research he had done on ID, prior to writing his attack on ID. (His piece, although passionate and clever in rhetoric, reflected a painful ignorance of the most basic facts of the ID debate.) "How many books on our side—the ID side—have you read?"

I asked, adding, "I would assume, of course, that you've read *Darwin's Black Box*, our most important book?"

His answer was short and truthful, "I haven't really read any books on the topic. I read a few essays." In a way, I was not surprised by his answer, but I was still shocked. "How unscholarly," I thought to myself, "for a college president to write a vehement, wholesale denunciation of ID without having done any serious homework on the topic." Since then, I have asked newspaper reporters, scientists, and entire audiences the same question: *Are ID bashers doing their homework? Who is taking time to read major works on both sides? Have we entered a period of runaway anti-intellectualism in the universities and media?*

The International Expansion of ID: On October 22, 2005, over seven hundred conferees from eighteen countries descended upon the glorious city of Prague, Czech Republic, to attend the first major conference on Intelligent Design ever held in Europe. Two "aha" moments were burned onto my memory that Saturday. One was when Stephen Meyer gave a talk on the DNA explosion that accompanied the Cambrian fossil explosion of so many complex phyla appearing on Earth. Paralleling the information in his notorious refereed journal article, it was probably the most riveting ID talk I have ever heard. Major naturalistic explanations for this DNA explosion were brought up one by one and were shown to fall short as explanations of the new genes needed for Cambrian morphology. It was a powerful line of argument, overwhelming in effect. Seated in the front corner, I was able to discreetly look back at the packed auditorium, gazing at the sea of seven hundred faces pondering intently Meyer's case for design. I could picture, like ripples spreading out from a rock dropped in a pond, this talk spreading through the European countries, setting off a European conversation over the Cambrian DNA mystery.

Another "aha" moment came later that day during a series of concluding talks by European speakers. One spellbinding presentation was a response by Dutch biophysicist and nanotechnology pioneer Cees (pronounced "Case") Dekker, a brilliant young scientist at the University of Delft. Dekker's research has been featured several times on the cover of the leading science journal, *Nature*. Although he is not personally doing ID lab research, he explained why he is in close correspondence with design theorists and why, in his opinion, the arguments and evidence for Intelligent Design theory had significant weight and held great promise. He showed why such research was especially needed in light of the severe difficulties that had come to light in the years since Michael Denton asked how much longer the Darwinian paradigm could last. Dekker, in a Sunday panel session, added that Denton's main points in *Evolution: A Theory in Crisis* still have never been adequately answered by Darwinists. He urged the audience to read Denton as one of the highest priority works for anyone exploring ID.

Cross-Examinations

Now that I've shared experiences that bolster confidence in my theses, it's time to subject them to a final cross-examination.

1. Isn't Intelligent Design an implausible line of thinking and religiously unnecessary in light of the brute fact that many Christians who are scientists, philosophers, and even theologians have spoken convincingly on behalf of Darwinian evolution as well-supported by the evidence and as "God's humble method" of creating? Those who ask this sort of question often cite John Haught, the Catholic theologian at Georgetown University who has written on God's display of humility and love in working through eons of time in creating life-forms. Haught argues that the God we meet in the Bible has a character by which he would not force himself on nature to coerce molecular matter to do what it would not normally do on its own. Others cite Francis Collins, the evangelical geneticist and director of the Human Genome Project, who has endorsed macroevolution as good science.

Still others have quoted, as a leader of "Evangelicals against ID," the Calvin College physicist (emeritus) Howard Van Till, who argues that God has richly gifted creation with the laws and substances to bring forth the diversity of life. (This later became known as the "Robust Formational Economy Principle" or RFEP for short.) Since the early 1990s, Van Till has been highly critical of Phillip Johnson and ID. He says that ID demeans God's excellence in creating by saying that he withholds certain gifts from creation and thus has to repeatedly step in, inserting information and monkeying with material systems to create new complexity. Even John Polkinghorne, the famous theoretical physicist turned Anglican clergyman at Oxford University, is brought out as one of this group, who is quite happy with the fine-tuned universe as evidence of the handiwork of God but who firmly declines to endorse the ID concept of God intervening since the creation of the universe. Several evangelical Protestant colleges are known to have biology professors who are adamantly opposed to ID as a promising theoretical framework. Are these examples, and many others that could be added, telling evidence against ID?

I think the first point to be made here (and it is one made by many ID theorists) is that all of these arguments are essentially theological in nature, or in a few cases, they simply reiterate the standard Darwinian scientific defenses without much engagement with the arguments from genetic complexity or irreducibly complex systems. Certainly, from the theological side there is nothing (except for insistence on a straight literal reading of time in Genesis and other biblical passages) to bar the Creator from employing secondary causes. In fact, when David says in Psalm 139:14–15, "I am fearfully and wonderfully made" and that his frame was "skillfully wrought" as if a fine tapestry, it is clear that the body of David was being produced through standard DNA replication and all the other biochemically understood processes of biological growth from zygote to adult.

The published ideas of Van Till (who recently announced his entry into process theology) and of Haught and Polkinghorne are expressed from a Christian presupposition that God was ultimately responsible for the universe, life, and mankind. They make for fascinating reading and function as hypothetical notions as to how and why God theoretically might have worked through secondary causes (as in Psalm 139) to accomplish his purposes of creating. Yet, the question is not, "Can this Darwinian framework be harmonized with a (general) Christian framework?" Clearly it can and has been, as various Protestant theological publications and many papal encyclicals and other statements from the Vatican have shown. The question rather, from a Christian's perspective, is, "Given God's purpose to create, why should I understand from the available evidence that God did in fact employ chance and necessity (which is the scientific description for "creation's giftedness" or creation's RFEP) to bring about highly complex systems and the 20,000 genes embedded in the genomes of higher animals?" In other words, it is in the scientific evidence, not in the realm of theological possibilities, where Christians should test the validity of scientific arguments.

2. Hasn't Behe's argument collapsed, especially since the Dover decision, and didn't that decision show, once and for all, that ID is "religion, not science"? This reflects the spirit of Kenneth Miller's triumphal rhetoric by which he claims to have vanquished the chief ID scientist by his book, *Finding Darwin's God*, the arguments of which are carried over with slight changes and additions in his testimony in the Dover trial in September and October 2005. Questions arising from the Dover decision could entail another entire chapter, but I defer the detailed discussion of Judge John E. Jones's decision to the new book *Traipsing into Evolution* and to the various critiques and interaction found on the Discovery.org website. One thing that has not escaped the notice of impartial observers is that the judge embraced—lock, stock, and barrel—every scientific argument set forward by the ACLU and NCSE witnesses (such as Pennock, Miller, Forrest, Gross, and others), and he seemed to have impatiently rejected the refutations of these scientific points made by Michael Behe or Scott Minnich and disregarded even the testimony of nontheist witness Steve Fuller, who testified that ID is "science, not religion."

In Appendix A of *Traipsing into Evolution* (published in March 2006 by the Discovery Institute) is a detailed reply of Michael Behe to the opinion of the court. I urge the reader to take time to read this piece, "Whether Intelligent Design Is Science," Behe's point-by-point response to Judge Jones's allegations made against the science of ID. Behe's response takes on twenty different points in just one section (the science section) of the opinion and sets out to refute the court on each of those twenty points. Like the other critiques published by the Discovery Institute, it shows the overreaching effect of the Dover decision—an announced attempt to forestall further legal action in such cases elsewhere in the U.S.—but in effect, Behe goes beyond that to demonstrate the factual errors embedded in the court's decision.

For example, the court finds that ID "violates the centuries-old ground rules of science by invoking and permitting supernatural causation." Behe replies, "It does no such thing. The Court's opinion ignores, both here and elsewhere, the distinction between an implication of a theory and the theory itself. As I testified, when it was first proposed the Big Bang theory struck many scientists as pointing to a supernatural cause. Yet it clearly is a scientific theory, because it is based entirely on physical data and logical inferences. The same is true of intelligent design."

Behe summarizes the court's missteps at the conclusion of his fourteen-page reply to the court:

> The Court's reasoning in section E-4 is premised on: a cramped view of science; the conflation of intelligent design with creationism; an incapacity to distinguish the implications of a theory from the theory itself; a failure to differentiate evolution from Darwinism; and strawman arguments against ID. The Court has accepted the most tendentious and shopworn excuses for Darwinism with great charity and impatiently dismissed evidence-based arguments for design. . . .
>
> All of that is regrettable, but in the end does not impact the realities of biology, which are not amenable to adjudication. On the day after the judge's opinion, December 21, 2005, as before, the cell is run by amazingly complex, functional machinery that in any other context would immediately be recognized as designed. On December 21, 2005, as before, there are no non-design explanations for the molecular machinery of life, only wishful speculations and Just-So stories.[10]

In other words, the court's decision cannot transform the scientific anomalies of the current paradigm into pluses for Darwinism—anomalies remain anomalies.

The ultimate vindication of Behe's scientific arguments will be fleshed out (I predict) in the decades to come. Meanwhile, to get a sense of why Behe's own confidence in his theory has steadily grown during its first ten years, one should carefully ponder his new Afterword, "Ten Years Later," in the tenth-anniversary edition of *Darwin's Black Box* (2006). This chapter should be mandatory reading for every educated citizen of planet Earth.

3. Isn't it true that ID is ultimately about moral issues—about God and about "red state vs. blue state" social issues—and not about scientific issues at all? I raise this question because it is one that is asked over and over, not only in the public media but even to me personally when speaking in front of college audiences. This was a question asked by a professor during the time for questions after my spring 2005 talk at Syracuse University, "Is There Evidence for Design?" I pointed out to the professor that not one single leader of ID was influenced by his or her religious worldview or conversion experience to embrace ID. Rather, in case after case, the facts show overwhelmingly that it was when they encountered powerful empirical critiques of Darwinism that they began to change their minds.

This is essentially the same question asked by Michael Ruse during my January 2006 debate, which took place before a packed auditorium at Valencia Community College in Orlando. I was astonished that Ruse would ask this "religious motivation" question, since it came after a twenty-five-minute opening in which he had presented virtually no scientific evidence to support the Darwinian paradigm. (He spent twelve minutes reviewing Darwin's life and showing slides that pictured microevolution of finches and tortoises on the Galapagos Islands. He briefly plugged *Archaeopteryx* as a transitional fossil between reptiles and birds and then spent the remaining twelve minutes talking about biblical religion and about silly statements made by Pat Robertson regarding the Dover decision.) When Ruse asked me, "This is all about gay marriage, isn't it?" I gave him and the audience essentially the same answer as I had given at Syracuse: the historical evidence documents over and over that ID theorists are being led to their views by the data, not dogma.

It is clear that religious issues are in play in this grand debate, but this is true just as much on the Darwin side as it is on ID's side. Many scholars working in the ID community have pointed out a key fact: Darwinism may not entail atheism, but it appears certain that to some extent, atheism entails Darwinism. Thus to question Darwinism on scientific grounds, to an atheist, cuts deep into the beliefs and presuppositions embedded in their own worldview. This is a dynamic fact of the ID-Darwinism debate that is often overlooked. It is clear also that ID theorists need to do a better job of explaining the interface between the question, "Is there real design in nature?" and the logical next question, "Who or what is the designer?" I am very interested in exploring the logical dynamics and the interdisciplinary nature of this crucial question. Clearly, those interested in ID do have deep interest in knowing the answer to the question: are there principled ways to proceed out of the bare design inference into investigating the nature and identity of the design-agent that will honor empirical evidence, logic, and respect for pluralistic rational discourse? I think there are, and much work remains to be done here.

4. Is not ID a betrayal of all we have gained in the Darwinian revolution, freeing science from the shackles of sectarian ideology? Darwin was the key figure in world history who cemented the transition to fully naturalistic science. Now ID is questioning this foundational assumption in science and thus is profoundly non- or anti-Darwinian in a philosophical sense. But there is another sense in which ID is *championing Darwin's positive legacy*—the legacy of *science as argument* as opposed to *science as truths learned by rote or by dogma or by selected exposure to evidence.* The latter phrase, with its three elements, epitomizes to me and to ID theorists the self-destructive educational agenda of Darwinian instruction today. I am convinced that Darwin would be horrified and deeply discouraged by the inversion of his intellectual legacy. After all, he opened a new phase of courageous, empirically based argumentation in science. In this regard, much to our surprise (or shock), Darwin's spirit of defending a scientific idea by presenting one long argument, which is incarnated by Intelligent Design rhetoric—is now striking back at his own heirs, the current twenty-first-century Darwinists. In the beginning of his *Origin of the Species,* he said, "I am well aware that scarcely a single point is discussed in this volume on which facts cannot be adduced, often apparently leading to conclusions directly opposite to those at which I have arrived. *A fair result can be obtained only by fully stating and balancing the facts and arguments on both sides of each question; and this cannot possibly be done here.*"[11]

If ID has served just one intellectual, educational, or scientific function in our age, it has been to awaken the scientific elite—and the science education industry in general—to the ongoing decay and impending collapse of the public plausibility of nature-driven macroevolution. Is this spreading doubt due to ignorance or religious dogma or worry about morality or dark theocratic motives? We are told that this is so almost daily by those desperate to stop the advance of Intelligent Design. But does the historical evidence support this picture of ID's motivation? Clearly, overwhelmingly, it does not. To the extent that responsible leaders of Darwinian science are engaging in (or silently tolerating) such motive mongering, they are just accelerating the collapse of their own paradigm. There are consequences for such egregious betrayal of truth and empirical reality. What ID is proposing to do—both in the intellectual world of public universities and in the public schools—is to allow empirical problems of Darwinism to be aired. They are carrying out the very program that Darwin urged in the quote above from his *Origin of the Species*. In addition, in forging a new paradigm that can test the phenomena attributable to the interaction of law and chance and that can test those only attributable to design, they are carrying out the same sort of daring project that Darwin himself initiated.

But more importantly, Darwin viewed science as *human dialectical argument*, and he set forth in his book one long argument for his theory. He faced his theory's problems and objections honestly, courageously, and forthrightly, with some amazingly creative flashes of rhetorical brilliance. I learned this from my mentor in rhetoric of science, John Angus Campbell, who is not only a world-renowned rhetorician of science but an equally renowned Darwin scholar. Campbell, who is famous for promoting dialogue and constructive engagement between the Darwinian and ID worlds, has taught us that Darwin himself, in his profound and unshakable commitment to proper scientific rhetoric, would be delighted to see ID engaging his view. He would give priority to vigorous and honorable dialogue, such as some leading Darwinists, including Michael Ruse, have been doing all along.

Would Darwin change his mind in light of the last thirty years of empirical critique since Denton, Johnson, and others launched ID and then Behe, Wells, Meyer, and Dembski have carried it to the next level? I think Darwin would have great pause at this point in history, in light of the complexity of the nano-wonders discovered inside living cells, along with their vast DNA libraries of digitized files. I think that he would be fascinated by the recent findings of microbiologist Ralph Seelke, with his "no more than two" discovery on the limit of mutations in nature. I think he would be attending every *OOF* seminar he could get to. But the importance is not what Darwin would conclude, but what would he say to Darwinists today who viciously distort ID, who create flamboyant and fearful nightmares about scientists leading the way to a theocracy, and who announce ID's threat to "ruin" the future of science. I think Darwin would pull aside his latter-day defenders and administer a grave rebuke. He would admonish them to stop distorting, to end the zero-concession policy, and to stop censoring teachers who show problems with his icons from peer-reviewed literature.

In *Doubts about Darwin*, I pointed out that the ultimate major player—truly the controlling player—in the drama of Darwin versus design is nature herself. She is stubborn, she is recalcitrant; *she is what she is*. And now she is speaking clearly to us with amazing messages about unexpected wonders at the root of biological life. If there is any lesson that Darwin, the meticulous empiricist, would leave for us at the height of our current drama, it is to be careful to not muffle or mythologize nature or to redescribe her "other than she really is." To violate this lesson is to run an intolerable risk. In Darwin's warning about nature, we find the ultimate act of *Darwin striking back*—striking against any dogma that would pronounce or pontificate on the reality of nature before she ascends to the witness stand. Nature is now speaking, her testimony is shocking and unexpected, and we need to listen.

Appendix

The following is an edited version of the opening section of Del Ratzsch's book review. The full review, nearly eight thousand words in length, is available at www.arsdisputandi.org/publish/articles/000191/article.pdf. It appeared in the June 2005 issue of the online journal *Ars Disputandi*.

How Not to Critique Intelligent Design Theory
A Review of Niall Shanks *God, The Devil, and Darwin* by Del Ratzsch

1 Introduction

[1]. . . [A]lthough I have argued elsewhere that some of the philosophical points made by a number of ID advocates are right, I have been critical of other aspects of ID views. . . . Having that interest, I would welcome a comprehensive, competent, evaluation and critique of ID. The structure, the catalogue of topics addressed, and the Oxford University Press imprimatur initially suggest that Niall Shanks's *God, the Devil, and Darwin*, may be exactly the book. . . .

[4] The book has its positive points. Various parts of the treatment of evolutionary theory and of thermodynamics are nice, wide-ranging discussions which may prove useful to some. . . .

[5] Unfortunately, however, this book seriously fails on crucial counts. Shanks has a substantive agenda (no surprise given that the "Foreword" is by Richard Dawkins, whose anti-religious emotionalism gets ever more shrill).

In his straining eagerness to denigrate anything associated with ID, Shanks inflates the rhetoric, misconstrues history, blurs important distinctions, and seriously skews the views of various ID advocates. And along the way there are repeated cries that the sky is falling. (For instance, although various critics argue that ID is a threat to science, education, Enlightenment values and so forth, were it not for Shanks it is unlikely that many of us would realize that the alleged progenitor of ID—creationism—is a threat even to NATO ('Introduction'), or that ID is in part really a cover for pushing religious extremist opposition to assisted suicide [p. 230].) . . .

[6] Overall this book is more likely to detract from than to contribute to objective and on-target discussion/evaluation/criticism of Intelligent Design. Consequently, I shall focus on what I take to be some of the major problems of the book. If ID and the ID movement do have serious flaws (and I will not here dispute that), then those should by all means be exposed rigorously and vigorously. But real exposure—or any sort of productive discussion—is not a likely immediate outcome of the sort of inaccuracies, slants and vilifications which unfortunately pervade this book.

2 A history of ID: skewing the past

[7] Critics of Intelligent Design routinely tar ID with a creationist brush ("Intelligent Design Creationism" is now the term of choice of ID critics), and although both polemically driven and in some sense misleading, use of the term is understandable given that significant numbers of lay creationists have enthusiastically appropriated ID into their own efforts. Nevertheless, the term is misleading because key figures in the birth and early development of the contemporary ID movement had no prior connection either with creationism or creationists. Key figures with no such prior ties include people like Phillip Johnson and the biochemist Michael Behe. . . . On the other hand, a number of dominant creationist figures have sharply criticized ID. That number includes Henry Morris. . . .

[8] Yet, Shanks simply asserts–without providing substantive evidence–that ID was "spawned" by the creationist movement [p. 6], which "gave rise to modern intelligent design theory." [p. 7] He further claims that "[m]odern biological creation science . . . descend[ed] with little modification from the positions articulated by Paley" [p. 35], and refers to "the natural theologians of old from whom they [modern creationists] descended" [pp. 48–9]. In this latter context, it is worth noting that (unless I missed it) in his definitive history of the creationist movement (*The Creationists*, University of California, 1993), Ronald Numbers does not so much as mention William Paley, *The Bridgewater Treatises*, the natural theology movement or other things which, if Shanks were right, would constitute the core roots of creationism.

Numbers traces contemporary creationism to the work of George McCready Price, and in the chapter devoted to Price the whole concept of design is mentioned only once in passing, and the design argument not at all. . . . Thomas Woodward . . . argues that the 1986 book *Evolution: A theory in crisis* by the Australian biochemist Michael Denton was an initial spur for both Johnson and Behe, and says that "It was Denton, more than anyone else, who triggered the birth of Design" (*Doubts about Darwin* . . . p. 32). Denton, who is generally identified as an agnostic during his entire adult life, is not a creationist by anyone's definition.

3 Ad hominem, distortion, and misrepresentation

[10] The level and type of *ad hominem* and otherwise derogatory rhetoric in this volume is really quite remarkable for something from Oxford University Press. We learn that ID advocates wish to ruin science, to close minds, that they engage in deception, they lie, they are actually bent on gaining political power for repressive and extremist purposes, etc. It often gets specifically personal. For instance, Shanks says that Phillip Johnson makes him think of people who "hang around schoolyards peddling soft drugs so that a taste for the harder stuff will follow" [p. 12].

[11] Attempts to denigrate frequently run to misrepresentation. Following are a number of examples. (I must ask [the] reader's indulgence for going on at some length—I think that the problem exhibited is serious and pervasive enough to warrant multiple, detailed examples. . . .)

[12] *A. William Dembski*. (Example 1) Shanks remarks, sarcastically, that Dembski (on p. 169 of his *No Free Lunch*, Rowman and Littlefield, 2002) "modestly claims to have discovered a fourth law of thermodynamics." Dembski's proposed fourth law is "something he calls the *Law of Conservation of Information*" [p. 123]. But 40 pages *prior* to the cited passage in *No Free Lunch*, Dembski says: "This is an instance of what Peter Medawar calls [in a 1984 book] the Law of Conservation of Information." [*No Free Lunch*, p. 129]

[13] Ten pages prior to the passage Shanks cites, Dembski says: ". . . Medawar's Law of Conservation of Information can therefore be formulated as follows . . ." [*No Free Lunch*, p. 159]

[14] As Dembski notes, there was discussion of a possible "Fourth Law" as early as the 1970s [*No Free Lunch*, p. 167]. Dembski simply suggested that Medawar's principle—which he repeatedly explicitly attributes to Medawar—is the law for which others had previously sought. That hardly fits Shanks's denigratory (and repeated) accusation.

[15] (One further oddity concerning Shanks's treatment of Dembski's work is worth noting. Despite supposedly presenting a scholarly study of

ID, Shanks does not so much as mention Dembski's initial scholarly ID book on complex specified information (*The Design Inference*, Cambridge University Press), published six years prior to Shanks's book. It is remarkable that a book subtitled "A Critique of Intelligent Design Theory" should completely overlook the theoretical manifesto of the design movement. On the contrary, much of what is cited in Shanks's discussion of Dembski comes from popularizations (e.g., an article with the subtitle "A Primer on the Discernment of Intelligent Design" in a collection from a popular Christian press, and another book from a different Christian publisher). This, like the above "history," is one of a number of instances of Shanks's apparently not having done the homework.)

Note: This review continues in the same vein of severe critique for another seventy numbered paragraphs.

Notes

Preface

1. The cover story, "Evolution Wars," was published in the August 15, 2005, issue of *Time*. I chatted with a contributing *Time* reporter who helped with this story, a young Princeton alumnus named Timothy Chu. He was covering for *Time* the ID conference in Greenville, South Carolina, that I was emceeing on August 5–6, 2005.

2. Paul Gross's quote is taken from the transcribed comments that were downloaded in early September 2005 from *The O'Reilly Factor* website under Radio & TV/The O'Reilly Factor Archive, and which were verified on the same website on January 22, 2006. See http://www.billoreilly.com/show?action=viewTVShow&showID=390#3. The italics are added for emphasis.

3. For more on the Dover trial and ID responses to Judge Jones's controversial decision, go to Discovery.org and see the links to articles and commentary on the Dover decision at the "Darwin Strikes Back" webpage. One of the most important is the February 3, 2006, analysis by Michael Behe, "Whether Intelligent Design Is Science," in which he levels twenty major criticisms of section E-4 (the science section) of the Dover opinion.

4. I'm referring here to the work of Ralph Seelke, at the University of Wisconsin (Superior), where his work on bacteria has revealed a "one mutation; two at the most" limit on what can be accomplished in bringing about new functional genes. See chapter 12 for further discussion of his work.

5. This paragraph, after the opening sentence, is using the form of argument developed by philosopher of science Stephen Meyer of the Discovery Institute.

6. Darwinism is often called "neo-Darwinism" to distinguish between the original ideas of natural selection acting on *random variation*—an idea focused more on normal variation among offspring rather than upon the input of genuine genetic novelty, which needs the event of genetic mutation to happen. Darwin knew very little of Mendelian genetics (he owned a copy of Mendel's work, but its pages were uncut!). The neo-Darwinian synthesis, which developed the model of *genetic mutation, sifted by natural selection*, was taking shape by the 1920s and was complete by the late 1940s. I use "Darwinism" in this book for "neo-Darwinism." This is standard practice.

7. My earlier book, a rhetorical history of Intelligent Design, included predecessors of ID in the 1960s and 1970s but mainly traced ID's birth and adolescence in the 1980s on through the year 2002. See Angus Menuge's chapter in *Debating Design: From Darwin to DNA*, ed. William A. Dembski and Michael Ruse (New York: Cambridge University Press, 2004) and two other excellent book-length

treatments: Denyse O'Leary, *By Design or By Chance?* (Minneapolis: Augsburg Books, 2004), and Larry Witham, *By Design: Science and the Search for God* (San Francisco: Encounter Books, 2003).

8. For a crash course on this field, read my appendix, "The Rhetoric of Science and Intelligent Design," in Thomas Woodward, *Doubts about Darwin* (Grand Rapids: Baker, 2003), 227–48.

9. Dawkins adds: "Or wicked—but I'd rather not consider that." More recently, Dawkins has affirmed this statement but has added another term to this list of explanations: brainwashed.

10. For a revealing historical sketch of Lewis's apparent change to a skeptic of macroevolution by the late 1950s, see the article "C. S. Lewis on Evolution: The Correspondence with Bernard Acworth," by Gary Ferngren and Ronald Numbers, posted at www.apologetics.org.

11. Michael Denton, *Evolution: A Theory in Crisis* (Bethesda, MD: Adler and Adler, 1986).

Chapter 1

1. The scolding of journalists came in the September 2005 issue of *Columbia Journalism Review*, in the cover story "Undoing Darwin," by Chris Mooney and Matthew Nisbet. Publisher Evan Cornog, an associate dean of the Columbia University Graduate School of Journalism, wrote a strongly anti-ID article in his August 21, 2005, column in *Media Nation*. The alleged danger to our civilization and its Enlightenment values is seen widely in anti-ID literature, but it dominates key chapters of the two Oxford University Press books: Niall Shanks, *God, the Devil, and Darwin: A Critique of Intelligent Design Theory* (Oxford, England; New York: Oxford University Press, 2004), and Barbara Forrest and Paul Gross, *Creationism's Trojan Horse: The Wedge of Intelligent Design* (Oxford, England; New York: Oxford University Press, 2004).

2. My word *painfully* is carefully chosen. The perception of the intrinsic danger of ID, and the public exposure of flaws in Darwinism, is a painful experience for Darwinists. Likewise, the resultant anti-ID rhetoric (often harsh, dismissive, and distorting) is painful to endure for those sympathetic to ID.

3. "Headline news" refers to the August 15 cover story of *Time*, after President Bush commented on ID during the August 1 meeting, and the August 14–16 front-page articles on ID in the *New York Times*. "Network newscasts" refers to many news programs that devoted time to the issue, including double coverage (twice in one month) on the *O'Reilly Factor*, a special *ABC Nightline* program, and an evening on *Larry King Live*.

4. See Ker Than's "Why Scientists Dismiss 'Intelligent Design,'" published on LiveScience.com, and posted on MSNBC.com on September 23, 2005.

5. Dr. Fritz Schaeffer, the pioneering founder of an entire field of chemistry, quantum computational chemistry (formerly at UC Berkeley, now at the University of Georgia); Dr. Cees Dekker of the University of Delft, a pioneer in biological nanotechnology; Dr. Fred Sigworth, an internationally known expert on physiology, teaching at Yale; Dr. Andrew Bocarsly, a leading chemical researcher at Princeton known for his work on inorganic materials that can be used in the conversion of sunlight into electrical current. This list could be extended greatly.

6. Denton, *Evolution*, 358.

7. Phillip Johnson's books, all published by InterVarsity (Downers Grove, IL), are: *Darwin on Trial* (rev. ed., 1993), *Reason in the Balance* (1995), *Defeating Darwinism by Opening Minds* (1997), *Objections Sustained* (1998), *The Wedge of Truth* (1999), and *The Right Questions* (2002). As of early 2006, all are still in print except *Objections Sustained.*

8. Charles B. Thaxton, Walter L. Bradley, and Roger L. Olsen, *The Mystery of Life's Origin: Reassessing Current Theories* (New York: Philosophical Library, 1984).

9. Richard Lewontin, "Billions and Billions of Demons," in the *New York Review of Books*, January 9, 1997, italics added.

10. Thomas Kuhn, *The Structure of Scientific Revolutions* (Chicago: University of Chicago Press, 1962).

11. Neo-Darwinism solidified in the 1940s when genetic mutations were incorporated as evolution's raw material, in lieu of natural variation. It essentially holds to "descent with modification of all living things through natural selection of random mutations and other natural mechanisms."

12. Letter from Bruce Alberts to NAS membership, April 2005.

13. See especially Michael Behe, "Irreducible Complexity: Obstacle to Darwinian Evolution," in *Debating Design: From Darwin to DNA*, ed. William A. Dembski and Michael Ruse (New York: Cambridge University Press, 2004).

Chapter 2

1. "Top Questions and Answers on Intelligent Design," September 9, 2005 (accessed October 6, 2005), Discovery.org. This is probably as simple and as official a definition as anyone could find (since it is published by the Discovery Institute).

2. This is the wording in the flyleaf of Niall Shanks's *God, the Devil, and Darwin*. The "dire threat" mentioned in this chapter is captured in the same flyleaf: "While ID has been proposed as a scientific alternative to evolutionary biology, Shanks argues that ID is in fact 'old creationist wine in new designer label bottles' and moreover is a serious threat to the scientific and democratic values that are our cultural and intellectual inheritance from the Enlightenment."

3. In chapter 4 I have an extended discussion on this point of Discovery Institute's stated goal of overthrowing the hegemony of the philosophy of naturalism. If needed, the reader can jump directly into that discussion for clarity on the simple point made here.

4. Michael Ruse, "Answering the Creationists: Where They Go Wrong and What They're Afraid Of," in *Free Inquiry* (1998).

5. Behe has made it clear that a "Supreme Being" cannot be implicated by the biochemical evidence. Ruse knows this. It is puzzling that he would expose himself to criticism based on a knowing distortion of simple facts.

6. Phillip Johnson has said often that this was one of the most popular jabs against his skepticism, especially in the interaction with Eugenie Scott and other staff at the National Center for Science Education. The charge against Behe—gross laziness—was observed just as widely, but notably in the reaction of Richard Dawkins (this was Dawkins's comment in a question and answer session in Berkeley attended by Phillip Johnson in 1997).

7. See Nancy Pearcey, "You Guys Lost!" in *Mere Creation*, ed. William A. Dembski (Downers Grove, IL: InterVarsity, 1998).

8. Stephen Jay Gould, *The Structure of Evolutionary Theory* (Cambridge, MA: Belknap Press, 2002). Gould's 1400-page book was completed just months before his death in May 2002. Not a line in the book mentions the ID movement or any of its theorists, even though Gould was well aware of the severe damage that ID already had done to public credibility of nature-driven macroevolution.

9. The one spot where Quammen seemed to cross over into macroevolution was his discussion of Gingerich's work on intermediate fossils leading to the first whales. On the *minimal complexity* question, see chapter 9, where recent estimates of the *minimal gene set* for the simplest cell are in the range of 250 genes to 1,000 genes or more.

10. See Michael Behe, *Darwin's Black Box* (New York: Free Press, 1996), chapter 1.

11. Behe says, in his book and his public lectures, that the big bang theory had clear religious implications, yet that did not prevent it from being taken seriously, and eventually embraced, by the scientific community.

12. The notion of naturalism as the all-important foundation of Darwinian confidence is clearly laid out in Phillip Johnson, *Darwin on Trial*, rev. ed. (Downers Grove, IL: InterVarsity, 1993). But it wasn't until Johnson's second book, *Reason in the Balance*, that he clarified the difference between metaphysical naturalism (a worldview) and methodological naturalism (a guideline used in scientific investigation). See Johnson's appendix in *Reason in the Balance*.

13. This assessment is based on a phone interview with one of the twenty authors of the book, whose name I have agreed to keep confidential.

14. Gerd B. Muller and Stuart A. Newman, eds., *Origination of Organismal Form: Beyond the Gene in Developmental and Evolutionary Biology* (Cambridge, MA: MIT Press, 2003), 3.

15. Ibid.

16. Ibid., 4, 7, italics added.

Chapter 3

1. See Woodward, *Doubts about Darwin*, chapter 4, in which I retold the story of this December 1989 meeting at the Campion Center on the west side of Boston.

2. Ibid., 83. I and all students of origins are eternally indebted to Dr. Raup for his courageous decision to allow these comments to see the light of day.

3. See ibid., 27, where I discuss some of the features of this review and its rhetorical significance.

4. For clarification on this journalistic garbling of the Pope's comments, see the text of the English translation of the communication to the Pontifical Academy, excerpted from the October 30, 1996, issue of the English edition of *L'Osservatore Romano*. This is on dozens of Internet websites (e.g., www.newadvent.org/library/docs_jp02tc.htm).

5. See Michael Behe, "Teach Evolution—and Ask Hard Questions," *New York Times*, August 17, 1999, and Michael Behe, "Design for Living," *New York Times*, February 7, 2005.

6. For the positive coverage in the *New York Times*, see Laurie Goodstein, "New Light for Creationism," *New York Times*, December 21, 1997, and James Glanz's story on ID on April 8, 2001. *Times* editorials on ID were consistently, and at times caustically, negative. See, for example, "Intelligent Design Derailed," *New York Times*, December 22, 2005.

7. See *Unlocking the Mystery of Life* (available through www.illustramedia.com), a documentary on ID where Behe makes this clear, as he does in many of his writings. My own discussions about Behe in *Doubts about Darwin*, especially in chapters 1, 7, and 8, show the essential role that Denton (and also Johnson) played in triggering his skepticism of naturalistic macroevolution.

8. Kenneth R. Miller, *Finding Darwin's God* (New York: Cliff Street Books, 1999).

9. Behe, of course, said that the evidence itself was silent on the specific identity or nature of the intelligence—even though Behe said he believed personally the designer was God.

10. See Dembski and Ruse, eds., *Debating Design*, 88, italics added.

11. Robert Pennock, *Tower of Babel* (Cambridge, MA: MIT Press, 1999), 37.

12. For an analysis of Pennock's critique of Phillip Johnson, see "Pennock vs. Johnson" at the "Darwin Strikes Back" webpage at Discovery.org/CSC.

13. Leonard Krishtalka, a Kansas paleontologist, called ID "creationism in a cheap tuxedo." This phrase exploded in popularity and soon began to circulate like a campaign slogan. See the closing paragraphs in the story on the state school board debate in Kansas, Pete Slevin, "Teachers, Scientists Vow to Fight Challenge to Evolution," *Washington Post*, May 5, 2005, A3.

14. All ID theorists were forceful on this point, but the most emphatic of all was Michael Behe in his many published comments. It is hard to find any work by Behe where he does not confront and dismantle the *creationist* tag.

15. Niles Eldredge, *The Triumph of Evolution and the Failure of Creationism* (New York: W. H. Freeman, 2000), 11. See also Niles Eldredge, *The Monkey Business: A Scientist Looks at Creationism* (New York: Washington Square Press, 1982).

16. Alan Linton, "Scant Search for the Maker," *Times Higher Education Supplement*, April 20, 2001, 29.

17. One was a volume he edited and contributed to, *Mere Creation*, based on papers presented at the conference by the same name held in November 1996 at Biola University. In 1998, he had published

a peer-reviewed but very technical book, *The Design Inference*, through Cambridge University Press. A year later, in 1999, Dembski added *Intelligent Design: The Bridge from Science to Theology*.

18. For details see Woodward, *Doubts about Darwin*, chapter 9.

19. William A. Dembski, *No Free Lunch: Why Specified Complexity Cannot Be Purchased Without Intelligence* (Lanham, MD: Rowman and Littlefield, 2002).

Chapter 4

1. In a few emails sent in 1994–1996, Phillip Johnson described his interaction with Carl Sagan. During his intense interaction with Johnson over dinner, Sagan said that a Christian's rationality (holding that God could act in the universe) is defective, whereas a rationality grounded upon naturalism was a superior and healthier form of reason.

2. The Darwinian "vindication" by the accumulation of "overwhelming evidence" was based on the blurring of the distinction between microevolution and macroevolution. ID theorists find this extrapolation from micro to macro illegitimate. Also, evidence for the ability of mutation-selection to drive such changes, or create new genetic information, was seen as nonexistent.

3. Daniel Dennett, *Darwin's Dangerous Idea: Evolution and the Meanings of Life* (New York: Simon & Schuster, 1995). Dennett said that Darwin's idea is a "Universal Acid" (the heading of the final pages of his book). He says, "Darwin's idea is a universal solvent capable of cutting right to the heart of everything in sight. . . . Some of the traditional details perish, and some of these are losses to be regretted, but good riddance to the rest of them." It is crystal clear from the book that any notion of a transcendent intelligence who played a detectable role in creating life and humanity is one of those ideas to which Dennett says "good riddance."

4. Theodosius Dobzhansky, "Nothing in Biology Makes Sense Except in the Light of Evolution," *American Biology Teacher* 35 (March 1973):125–29.

5. Forrest and Gross, *Creationism's Trojan Horse*. See my listing and response to these distortions in the article that can be linked from the *Darwin Strikes Back* webpage at Discovery.org/CSC.

6. A sampler of these, listed at www.NCSEweb.org on December 10, 2005, are resolutions from the Biophysical Society, the American Association of University Professors, and the American Chemical Society. Also, the American Astronomical Society passed a resolution affirming evolution and denouncing ID, after a similar joint resolution by the American Society of Agronomy, the Crop Science Society of America, and the Soil Science Society of America.

7. Dan Peterson, "What's the Big Deal About Intelligent Design?" *The American Spectator* (December 2005/January 2006). Peterson had authored an earlier, brilliant piece on ID in the same magazine in the July 2005 issue.

8. In my own televised encounter with two critics of ID on Kathy Fountain's midday *My Turn* program in Tampa in September 2005, Eddie Tabash described Dembski as a "fundamentalist." I quickly challenged this as inaccurate.

9. I have heard about ten of Behe's lectures given in different conferences, and after 1998 he almost always included this "religious motivation" discussion.

10. Dembski and Ruse, eds., *Debating Design*, 329.

11. Cornell's interim president, Hunter Rawlings III, used H. Allen Orr's article attacking ID, in the May 30, 2005, issue of the *New Yorker*, as one of his key sources when he gave his "State of the University" speech against ID on October 21, 2005. The entire text, which I have read and studied carefully, was printed the next day in the *Ithaca Journal*. President Timothy P. White of the University of Idaho, meanwhile, released his statement in November 2005, according to the report in "ARN-Announce," no. 50, December 1, 2005, published by the Access Research Network, ARN.org. ID micro biologist Scott Minnich teaches at the University of Idaho; his views may have prompted this edict.

12. Shanks, *God, the Devil, and Darwin*. He uses the term *extremists* (and the cognate *extreme*) eleven times in his brief preface and the first chapter as his favorite synonym for creationists in general. Especially notable is the number of times (such as on p. xi, most importantly) that this term is referring

to the "ID creationists." On pages 224–26 there are fourteen uses of *supernatural* or a cognate. The same is true of pages 15–18, where *supernatural* appears twelve times.

13. Ibid., xi. The first italics are mine, and the second italics are in the original.

14. See Del Ratzsch's review, "How Not to Critique Intelligent Design Theory," in the 2005 issue of the online journal *Ars Disputandi*. This reivew, reprinted in part in the appendix, can be accessed at www.arsdisputandi.org/publish/articles/000191/article.pdf. Neil Manson's review of Shanks was published on May 9, 2004 in the online journal *Notre Dame Philosophy Reviews*. It was accessed at http://ndpr.nd.edu/review.cfm?id=1437.

15. Accessed October 2005, the article was found on APS News Online on the webpage devoted to the regular column called "The Back Page," www.aps.org/apsnews.

16. These excerpts and comments are taken from a Truth Sheet 03-05 (revised 7/05), which is entitled: "The 'Wedge Document" How Darwinist Paranoia Fueled an Urban Legend." It, along with the much more extensive document, "The 'Wedge Document": 'So What?'" were both downloaded from the Center for Science and Culture website (a subset of Discovery.org) on December 13, 2005.

17. All quotes are from the one-page document, Truth Sheet 05-01, entitled "Discovery Institute and 'Theocracy,'" downloaded on December 13, 2005.

18. See not only *The Wedge of Truth* but also *Defeating Darwinism by Opening Minds* and Johnson's first chapter of *Signs of Intelligence: Understanding Intelligent Design*, ed. William A. Dembski and James M. Kushiner (Grand Rapids: Brazos Press, 2001).

19. Kuhn, *The Structure of Scientific Revolutions*.

20. The italicized words are my attempt to summarize the thrust of this crucial chapter in Denton's *Evolution*.

21. William A. Dembski, *The Design Revolution: Answering the Toughest Questions about Intelligent Design* (Downers Grove, IL: InterVarsity, 2004), 21.

22. Mark Perakh, *Unintelligent Design* (Amherst, NY: Prometheus Books, 2004); Matt Young and Taner Edis, eds., *Why Intelligent Design Fails: A Scientific Critique of the New Creationism* (New Brunswick, NJ: Rutgers University Press, 2004).

23. John A. Campbell and Stephen C. Meyer, eds., *Darwinism, Design, and Public Education* (East Lansing, MI: Michigan State University Press, 2003).

24. Ronald Numbers, *The Creationists* (Berkeley: University of California Press, 1993).

25. *Unlocking the Mystery of Life* and *The Privileged Planet*, Illustra Media, 2002 and 2004. The DVDs can be ordered through www.illustramedia.com.

26. Guillermo Gonzalez and Jay Richards, *The Privileged Planet: How Our Place in the Cosmos Is Designed for Discovery* (Washington, DC: Regnery Publishing, 2004).

27. Personal interviews via phone and in person with graphics artist Tim Doherty, October–December 2002 and January 2003.

28. *Traipsing into Evolution* was published by Discovery Institute Press and written by David K. DeWolf, professor of law at Gonzaga University, Dr. John G. West, associate professor and chair of the political science department at Seattle Pacific University, Casey Luskin, attorney and program officer for public policy and legal affairs at Discovery Institute, and Dr. Jonathan Witt, senior fellow and writer in residence at Discovery Institute.

29. See Edward B. Daeschler, Neil H. Shubin, and Farish A. Jenkins Jr., "A Devonian Tetrapod-like Fish and the Evolution of the Tetrapod Body Plan," *Nature* (April 6, 2006); and Neil H. Shubin, Edward B. Daeschler, and Farish A. Jenkins Jr., "The Pectoral Fin of *Tiktaalik Roseae* and the Origin of the Tetrapod Limb," *Nature* (April 6, 2006).

30. The quote is from discovery.org/csc, in "Irreducible Complexity Stands Up to Biologist's Research Efforts," by Discovery Institute staff, posted April 6, 2006. The *Science* article was written by Jamie Bridgham, Sean Carroll, and Joe Thornton.

31. The Behe and Meyer quotes are from the same posted article ("Irreducible Complexity Stands Up"), which itself included some quotes from Michael Behe's article posted April 6, 2006, on idthefu-

ture.com, entitled, "The Lamest Attempt Yet to Answer the Challenge Irreducible Complexity Poses for Darwinian Evolution."

Chapter 5

1. See, for example, the appreciative remark by David Ussery about Behe's research on Z-DNA in Young and Edis, eds., *Why Intelligent Design Fails*, 48.

2. Lee Strobel, *The Case for a Creator* (Grand Rapids: Zondervan, 2004), 195.

3. This was the comment made at a Festschrift dinner in honor of Phillip Johnson at Biola University in Los Angeles in April 2004.

4. Shanks, *God, the Devil, and Darwin*, 160, italics added. In the final line on Behe's role in ID, Shanks may have intended to compare him to Plato's role in European philosophy. A. N. Whitehead said the European philosophical tradition "consists of a series of footnotes to Plato." This is found in Alfred North Whitehead, *Process and Reality: An Essay in Cosmology* (New York: Free Press, 1979), 39.

5. Shanks, *God, the Devil, and Darwin*, 164.

6. "Top Questions," accessed December 29, 2005, Discovery.org/CSC.

7. Darwin's famous "If it could be demonstrated . . . " quote was also used by Michael Denton in *Evolution: A Theory in Crisis*, and Phillip Johnson in *Darwin on Trial*.

8. Behe, *Darwin's Black Box*, 39, which Behe's endnote references Charles Darwin, *Origin of the Species* (Washington Square, NY: New York University Press, 1988), 154. It can also be found in Charles Darwin, *Origin of the Species* (New York: Bantam Classics, 1999), 158.

9. Darwin, *Origin of the Species*, Bantam Classics, 161–62.

10. Ibid., 158.

11. Perakh, *Unintelligent Design*, 118–19.

12. Darwin, *Origin of the Species*, Bantam Classics, 161–62.

13. For an extension of this chapter, the reader is invited to visit Discovery.org, and access the essay "Irreducible Complexity on Trial, 1996–2006 through the "Darwin Strikes Back" webpage. Also go to ARN.org where there are seven articles extending and applying Behe's arguments and eleven lengthy reply essays, responding to critics over the years.

14. For those new to irreducible complexity: The mousetrap models IC because it couldn't evolve step by step. You don't catch mice unless all five parts are in place and well matched: base, U-shaped hammer, spring, holding bar, and sensitive catch. Take one part away, and no mouse is caught, showing the trap's IC.

15. See Behe's response to Keith Robison in "Behe Responds to Postings in Talk Origins Newsgroup," accessed January 10, 2006, at http://arn.org/docs/behe/mb_toresp.htm.

16. See http://udel.edu/~mcdonald/mousetrap.html.

17. See www.millerandlevine.com/km/evol/DI/Mousetrap.html.

18. Dembski and Ruse, eds., *Debating Design*, 366.

19. ID theorist William Lane Craig attended a national philosophical meeting and heard emphatic acceptance of this criticism of Behe's analogy as a refutation. He was astonished that clearheaded philosophers would be so muddled. It is a breach of logic to say that a flaw in a teaching analogy indicates a collapse of an argument. Craig said that any scientist could have a robust theory but use poor analogies to capture the scientific reality. Debunking an analogy in no way debunks the theory. William Lane Craig, Trinity College, Trinity, Florida, April 2002.

20. Darwinists claim that IC can evolve, pictured in an analogy of a stone arch. If we see a stone arch, with blocks resting on each other, we assume special equipment was needed to suspend blocks, pending the keystone being dropped into place. Some (including Michael Ruse, "Where the Creationists Go Wrong," *Free Inquiry*, 1998) say that one could have a dirt mound upon which blocks came to rest, lined up from one side to the other. If the dirt was washed away, a complex arch is left! The additional part (a mound) facilitated the joining of the blocks, but it was removed after the blocks were in place.

A cellular IC system might evolve the same way, with additional parts or stages lost along the way. Ruse used this analogy in our debate in January 2006.

This analogy has many flaws: (1) The next-to-final arrangement (before dirt removal) is more complex than the final arrangement. What Darwinian process led to this greater complexity? (2) In the "arch-on-mound" analogy, the conglomeration has no useful function; its complexity lacks a plausible function-driven Darwinian explanation. (3) What physically lined and fitted the blocks? (4) Most importantly, the crudity of the analogy (also inherent in the mousetrap) is that individual parts (blocks, dirt mound) hardly capture the precise shape and incredible improbability of even a single protein arising through a random process. The precise folded shape is key; it must work in concert with other proteins. Arch stones are crude things compared to proteins. Perhaps an arch may form this way once in a blue moon. But what are the probabilities of any *single protein*, in the flagellum set of forty, forming through the shuffling processes of a living cell? Arch analogies are exercises in irrelevance.

21. Almost all of the early essays of Behe in response to his critics made the point that biochemists reviewing his book almost universally admitted that pathways to IC were unknown, but hopefully would be eventually discovered. See the eleven essays listed on Behe's home page at ARN.org.

22. When Dawkins visited a bookstore in Berkeley, California, in the 1996–1997 timeframe, to push his *Climbing Mount Improbable*, Phillip Johnson sat in the front row at his talk. During the question period, Johnson asked Dawkins his view on Behe, whereupon Dawkins made the "lazy" charge. This is based on email and personal conversations around that time. Other eyewitnesses have heard him say the same thing.

23. This was accessed on 17 May 2006 at www.iscid.org/papers/Dembski_StillSpinning_030403.pdf, and is found on page 1 of the PDF document.

24. Perakh, *Unintelligent Design*, 117.

25. Ibid., 118, italics added.

26. Quoted by Dembski, "Eliminative Induction," in *The Design Revolution*, 220.

27. For this wording I am indebted to John Warwick Montgomery, who used this term in a transcribed discussion, which comprises the appendix of his slender classic, *History and Christianity* (Downers Grove, IL: InterVarsity, 1967).

28. Dembski, *The Design Revolution*, 221.

29. Ibid.

30. "A True Acid Test: Response to Ken Miller," posted on July 31, 2000, Discovery.org. http://www.discovery.org/scripts/viewDB/index.php?command=view&id=441.

31. Russ Doolittle, "Delicate Balance," *Boston Review*, February/March 1997.

32. For an updated version of this article, see Scott Minnich, "Genetic Analysis of Coordinate Flagellar and Type III Regulatory Circuits in Pathogenic Bacteria," chapter 13 in *Darwin's Nemesis*, ed. William Dembski (Downers Grove, IL: InterVarsity, 2006).

33. Such folding processes, leading to a complex 3-D conformation, involves in some cases the assistance of special barrel-shaped machines, in which special chaperone or chaperonin proteins assist in folding.

34. See a description of Doug Axe's latest work in the peer-reviewed article on the Cambrian information explosion by Stephen Meyer, "The Origin of Biological Information and the Higher Taxonomic Categories," *Proceedings of the Biological Society of Washington* 117, no. 2 (August 4, 2004): 213–39.

35. See Michael Behe's chapter in *Darwinism: Science or Philosophy?* ed. Jon Buell and Virginia Hearn (Richardson, TX: Foundation for Thought and Ethics, 1993).

Chapter 6

1. In "Critics Rave over *Icons of Evolution*: A Response to Published Reviews," www.discovery.org, June 12, 2002, Jonathan Wells says: "One egregious example of Darwinian censorship occurred in 2000 and 2001 in Burlington, Washington. High school biology teacher Roger DeHart tried to supplement

his biology textbook with articles critical of Haeckel's embryos and peppered moths from mainstream science publications such as *The American Biology Teacher*, *Natural History*, *The Scientist*, and *Nature*. The American Civil Liberties Union issued veiled threats of legal action, and the National Center for Science Education, a pro-Darwin advocacy group with which reviewers Scott, Padian, and Gishlick are all affiliated, insisted that DeHart teach only orthodox Darwinism. Bowing to the intimidation, the superintendent of DeHart's school district prohibited him from distributing the articles—or even talking about them! DeHart was subsequently removed from his biology teaching position." (This episode is well documented in the video *Icons of Evolution*.)

2. Wells, "Critics Rave over *Icons of Evolution*," 1.

3. Kevin Padian and Allan Gishlick, "The Talented Mr. Wells," *The Quarterly Review of Biology* (March 2002).

4. Larabell's letter, quoted in "Critics Rave over *Icons of Evolution*."

5. Quotations regarding all ten icons are taken from Forrest and Gross, *Creationism's Trojan Horse*, 99–100.

6. See Jonathan Wells, *Icons of Evolution: Science or Myth? Why Much of What We Teach about Evolution Is Wrong* (Washington, DC: Regnery Publishing, 2000), 231–35, for his discussion of possible fraud in biology textbooks.

7. Ibid., 251.

8. Ibid., 51.

9. My father, William W. Woodward, studied evolution at Princeton, where he received his A.B. degree in 1928. His belief in macroevolution faded near the end of his life. Before his death in 1992, he was a budding fan of Phillip Johnson, having met him personally at dinner and having enjoyed hearing him on William F. Buckley's *Firing Line* in 1991.

10. Wells, *Icons of Evolution*, 91.

11. Wells, "Critics Rave over *Icons of Evolution*." His endnote provides this set of reference data: Charles Darwin, *The Origin of the Species*, chapter 14; *The Descent of Man*, chapter 1. The quotation calling embryology "by far the strongest" evidence is from a September 10, 1860, letter to Asa Gray, in Francis Darwin, ed., *The Life and Letters of Charles Darwin*, vol. 2 (New York: D. Appleton, 1896), 131; the letter is cited in Ernst Mayr, *The Growth of Biological Thought* (Cambridge: Harvard University Press, 1982), 470, and in Stephen Jay Gould, *Ontogeny and Phylogeny* (Cambridge, MA: Harvard University Press, 1977), 70.

12. The data referenced here, along with Wells's quote, are from Wells, *Icons of Evolution*, 82–83.

13. Ibid., 91.

14. Ibid. is the source for both quotes. In the first Richardson quote, the words are Richardson's; the second quote is Wells's summary of the results.

15. Ibid., 92.

16. Wells writes: "Although von Baer accepted the possibility of limited transformation of species at lower levels of the biological hierarchy, he saw no evidence for the large-scale transformations proposed by Darwin. For example, von Baer did not believe that the various classes of vertebrates . . . were descended from a common ancestor." Ibid., 85.

17. This tape is in my possession. The debate, in front of over a thousand attendees, was at Dillon Gymnasium, Princeton University, Princeton, NJ, in April 1980.

18. See the section entitled "Resurrecting Recapitulation," in Wells, *Icons of Evolution*, 88–90.

19. Wells develops this three-point outline from Dawkins's advice, "It is absolutely safe to say that if you meet somebody who claims not to believe in evolution, that person is ignorant, stupid or insane (or wicked, but I'd rather not consider that)." See Richard Dawkins, "Put Your Money on Evolution," *New York Times*, April 9, 1989, section 7, 35.

20. Wells, "Critics Rave over *Icons of Evolution*."

21. Jerry Coyne, "Creationism by Stealth," *Nature* 410 (April 12, 2001): 745–46. This quote is from p. 745.

22. The following block quote paragraphs are all from Wells, "Critics Rave over *Icons of Evolution*."

23. Source information provided in Wells, "Critics Rave over *Icons of Evolution*," is: Adam Sedgwick, "On the Law of Development Commonly Known as Von Baer's Law; and on the Significance of Ancestral Rudiments in Embryonic Development," *Quarterly Journal of Microscopical Science* 36 (1894): 35–52.

24. Source information provided in Wells, "Critics Rave over *Icons of Evolution*," is: William W. Ballard, "Problems of Gastrulation: Real and Verbal," *BioScience* 26 (1976): 36–39, 38; Richard P. Elinson, "Change in Developmental Patterns: Embryos of Amphibians with Large Eggs," in R. A. Raff and E. C. Raff, eds., *Development as an Evolutionary Process*, vol. 8 (New York: Alan R. Liss, 1987), 1–21, quote from p. 3. See also Jonathan Wells, "Haeckel's Embryos and Evolution: Setting the Record Straight," *The American Biology Teacher* 61 (1999): 345–49.

25. Source information provided in Wells, "Critics Rave over *Icons of Evolution*," is: Coyne, "Creationism by Stealth," 745.

26. Wells, "Critics Rave over *Icons of Evolution*."

27. Scott's review is quoted from Wells, "Critics Rave over *Icons of Evolution*." The original review is "Fatally Flawed Iconoclasm," *Science* 292 (June 22, 2001): 2257–58.

28. Eugenie C. Scott, "Fatally Flawed Iconoclasm," *Science* 292 (June 22, 2001): 2257–58.

29. Wells, "Critics Rave over *Icons of Evolution*."

30. Ibid.

Chapter 7

1. "Nonexistent" could be rephrased "virtually nonexistent." At higher taxonomic levels (kingdoms, phyla, and classes), "virtually" should definitely be dropped.

2. See Simon Conway-Morris's chapter on the Cambrian in Muller and Newman, eds., *Origination of Organismal Form*.

3. From chapter 4, "Natural Selection," in Darwin, *Origin of the Species*, Bantam Classics, 71.

4. See Darwin's comment, ibid., 169: "We have seen that species at any one period . . . are not linked together by a multitude of intermediate gradations . . . partly because the very process of natural selection almost implies the continual supplanting of preceding and intermediate gradations." David Raup told me in the fall of 2000 that Darwin's view on competitive replacement was simply wrong.

5. From the video *Icons of Evolution*, available through www.coldwatermedia.com.

6. See Wells, *Icons of Evolution*, 57, where he discusses the controversial ideas of Harry Whittington and Malcolm Gordon. Both doubted the monophyletic (single trunk of the tree of life) point of view.

7. In the *Icons of Evolution* video, high school biology teacher Roger DeHart explains that in his school's biology textbook there is one sentence about the Cambrian that merely mentions the explosion and moves on to blithely declare the tree of life (and the developments of evolution as powered by natural selection) to be a fact.

8. Donald Prothero, "The Fossils Say Yes," *Natural History* (November 2005): 52–56. Prothero's title is possibly inspired by the original title of a popular creationist book by Duane Gish, *Evolution: The Fossils Say No!*

9. Stephen Jay Gould, *Wonderful Life: The Burgess Shale and the Nature of History* (New York: W. W. Norton, 1989); Simon Conway-Morris, *The Crucible of Creation: The Burgess Shale and the Rise of Animals* (New York: Oxford University Press, 1998). I also recommend the outstanding interaction essay (a conversation in print) between Conway-Morris and Gould after Conway-Morris published *Crucible of Creation* in 1998. Entitled "Showdown on the Burgess Shale," it is available at www.stephenjaygould.org/library/naturalhistory_cambrian.html. It is mandatory reading for anyone digging into the Cambrian controversy.

10. Stephen Jay Gould, "Treasures in a Taxonomic Wastebasket," *Natural History*, December, 1985.

11. Darwin, *Origin of the Species*, Bantam Classics, 230.

12. The senior Cambridge professor Harry Whittington did the initial work, and his two Ph.D. students, Derek Briggs and Simon Conway-Morris, followed on and were able to make a number of the most spectacular discoveries.

13. Of course, the body plans of land plants appear later, and several tiny life-forms, mainly parasitic types of phyla, are not evident in the Cambrian as of this time.

14. Miller, *Finding Darwin's God*, 124–25.

Chapter 8

1. In case any Ohioans are curious, the small farm town was Canal Winchester.

2. See ISSOL.org, where summaries of triennial meetings of ISSOL are published.

3. Oparin had the set of four gases that Stanley Miller used, including two more not mentioned here: methane and hydrogen. Haldane suggested carbon dioxide as a source of carbon (instead of methane). For a helpful comparison of the two, see www.daviddarling.info/encyclopedia/O/OparinHaldane.html; this was my source of information.

4. This sentence has a generality that scientists deride as "hand-waving." But sometimes, in developing a new theory, such generalities are simply unavoidable at the earliest stages of theory building.

5. One report at the 1999 ISSOL conference said that, almost assuredly, the early earth had an atmosphere composed of water, carbon dioxide, and nitrogen gas. This research finding was a significant disappointment. This report is by Fazale Rana and Hugh Ross, "An Inside Report on ISSOL '99: Life from the Heavens? Not This Way . . . ," Reasons.org, accessed on December 29, 2005.

6. A. E. Wilder-Smith's classic, *The Natural Sciences Know Nothing of Evolution,* is a superb introduction to the topic of chirality in proteins.

7. One amino acid, glycine, has no stereospecificity; it does not possess right-handed and left-handed variations.

8. This problem is mentioned in Paul Davies, *The Fifth Miracle: The Search for the Origin and Meaning of Life* (New York: Simon and Schuster, 1999); it is made clear that no known solution seems to be forthcoming. This is typical; the same is true for every other book I consulted, including Lahav's work (see following note).

9. Noam Lahav comments on the "progenote": "But first let us turn to the progenote—the hypothetical creature which is on the verge of the realm of extant biology—and use it as a departure point in our back-extrapolated voyage toward the very beginning of biology." Noam Lahav, *Biogenesis: Theories of Life's Origin* (New York: Oxford University Press, 1999), 143.

10. Charles B. Thaxton, Walter L. Bradley, and Roger L. Olsen, *The Mystery of Life's Origin: Reassessing Current Theories* (New York: Philosophical Library, 1984).

11. See Woodward, *Doubts about Darwin*, 85–91, on Thaxton and the impact of *Mystery*, including a sampler of the explosive reactions it received from scientists, including Yale biophysicist Harold Morowitz, who wrote a positive review.

12. Dean Kenyon and Gary Steinman, *Biochemical Predestination* (New York: McGraw-Hill, 1969).

13. The primary mechanism, "photodissociation," was virtually unavoidable: water molecules (H_2O) high in the atmosphere, when exposed to ultraviolet radiation, would have the hydrogen atoms knocked loose from the oxygen. The oxygen would blend in with Earth's atmosphere.

14. The separating out of the different kinds of entropy work (chemical, thermal, and configurational) became part of the ongoing argument for design in the chemical evolution field, and it was a founding part of the intellectual structure of the ID Movement.

15. The five other ideas besides naturalistic abiogenesis are: (1) new natural laws, (2) panspermia, (3) directed panspermia, (4) creator within the cosmos, (5) creator outside the cosmos.

16. James Jekel, review of *The Mystery of Life's Origin*, *Yale Journal of Biology and Medicine* (December 1984); Klaus Dose, "The Origin of Life: More Questions Than Answers," *Interdisciplinary Science Reviews*, vol. 13, no. 4, 348.

17. Robert Shapiro, *Origins: A Skeptic's Guide to the Creation of Life on Earth* (New York: Bantam Paperback, 1986). This is one of the finest books ever written on chemical evolution, and the dazzling dust jacket blurbs that he was able to garner speak for themselves and are the basis for my description of the book as "celebrated."

18. Robert Shapiro, "A Replicator Was Not Involved in the Origin of Life," in *IUBMB Life* (A Journal of the International Union of Biochemistry and Molecular Biology) 49 (2000): 173–75; cited in Walter Bradley's chapter, "Information, Entropy, and the Origin of Life," in *Debating Design*, ed. Dembski and Ruse (Cambridge University Press, 2004), 346.

19. See de Duve's own autobiography page at: nobelprize.org/medicine/laureates/1974/duve-autobio.html. I have drawn much of my material from this webpage, accessed on December 28, 2005. His role as participant in ID-type discussions is retold in Phillip Johnson's chapter, "The Information Quandary," in *The Wedge of Truth*, and in my discussion of "The Nature of Nature" conference hosted by Dembski at Baylor University, discussed in chapter 9 of Woodward, *Doubts about Darwin*.

20. Thioesters are energy-rich metabolites, and for a good summary of this theoretical phase, see Lahav, *Biogenesis*, 262–63.

21. See Fazale Rana and Hugh Ross, *Origins of Life: Biblical and Evolutionary Models Face Off* (Colorado Springs: NavPress, 2004), 52. This view is so important to Lahav that Wächtershäuser's work is cited on a total of about fifty different pages in a text of three hundred pages—one in six pages discusses this theory!

22. A. G. Cairns-Smith is best known for his book *Seven Clues to the Origin of Life: A Scientific Detective Story* (New York: Cambridge University Press, 1985), but he earlier published *Genetic Takeover and the Mineral Origins of Life* (New York: Cambridge University Press, 1982). Wikipedia has a wonderful and helpful summary of his clay theory at http://en.wikipedia.org/wiki/Graham_Cairns-Smith.

23. Francis Crick, *Life Itself: Its Origin and Nature* (New York: Simon and Schuster, 1981). In *Mystery of Life's Origin*, Bradley, Olsen, and Thaxton discussed panspermia in their epilogue. Paul Davies, in his key book *The Fifth Miracle*, devotes an entire chapter (23 pages) to this idea.

Chapter 9

1. See the extremely valuable summary of this point in chapter 12 of Rana and Ross, *Origins of Life*.

2. These facts, although presented by speakers at the Greenville Conference, are here derived from Rana and Ross, *Origins of Life*. The E coli figure is on p. 165 and the independent-cell figure (1500–1900) is on p. 162.

3. For this vivid picture of such parasitic life, I am indebted to the descriptions given by several speakers at the "Uncommon Dissent" forum on ID in Greenville, SC, in August 2005.

4. See Rana and Ross, *Origins of Life*, 163, and their referenced sources in the cited endnotes for chapter 12 (especially endnotes 11 through 14).

5. Edward Peltzer especially gave some credit to researchers for modest results at this stage.

6. Kenyon and Steinman, *Biochemical Predestination*.

7. *Unlocking the Mystery of Life*, Illustra Media, 2002. The DVD can be ordered through www.illustramedia.com.

8. Ibid.

9. Ibid.

10. See "The Information Quandary" in Johnson, *The Wedge of Truth*.

11. Davies, *The Fifth Miracle*, 270.

12. For an expansion on this critical analysis of naturalism's explanatory dilemma, see the opening chapters of C. S. Lewis, *Miracles: A Preliminary Study* (1947; reprint, New York: Macmillan, 1978).

13. Davies, *The Fifth Miracle*, 246.

14. Ibid., 257–58.

15. Ibid., 259, italics added.

16. Gould, *Structure of Evolutionary Theory*. See the footnote on page 101 in which Gould lists the creationist syllogism.

17. This dodging strategy, with a closing comment of affirmation of progress, seems to be the approach taken in Forrest and Gross, *Creationism's Trojan Horse*, 99–103.

18. Lahav, *Biogenesis*, 303. His reference note is to Arrhenius et al., "Entropy and Charge in Molecular Evolution—the Case of Phosphate," *Journal of Theoretical Biology* 187 (1997): 503–33.

19. Lahav, *Biogenesis*, 303–4, italics added.

20. Ibid.

21. Walter Bradley, "Information, Entropy, and the Origin of Life," in *Debating Design*, ed. Dembski and Ruse, 347–48.

22. Ibid.

23. In *Doubts about Darwin*, I pictured Denton's skeptical critique of macroevolution as this same imaginary situation of a scientist looking for a way out of an underground labyrinth. See Woodward, *Doubts about Darwin*, 256 n. 18.

24. See "Science vs. Religion, to Science vs. Science," Montana News Association, http://www.montanasnews.com/article.php?molde=view&id=2677, accessed on March 16, 2006.

Chapter 10

1. *The Day the Earth Stood Still* (1951) was directed by Robert Wise (known for *The Sound of Music*, *Star Trek: The Movie*, and dozens of other credits). The short plot summary that follows is adapted from www.imdb.com/title/tt0043456—IMDb's website (which touts itself as the world's largest movie database), accessed on December 30, 2005.

2. Of course, for those who insist on precision, I'm leaving out the start and stop codons at the beginning and end of the other one hundred codons, so technically, there are at least 102 codons in the entire string.

3. The work of Ralph Seelke at the University of Wisconsin has focused on this question. It appears natural selection has a three-mutation limit for the formation of new meaningful DNA sequences. See chapter 12 for details.

4. See Stephen Meyer's famous review article, "The Origin of Biological Information and the Higher Taxonomic Categories." The article can be viewed online at Discovery.org. Meyer references E. Koonin's "How Many Genes Can Make a Cell?" *Annual Review of Genomics and Human Genetics* 1 (2000): 99–116.

5. Of course I'm not even counting the non-coding regions, which in the past has been called "junk DNA"—of higher eukaryotes. Recent research seems to be indicating that this "junk DNA" is not so junky after all, and it may have a number of purposes that were not glimpsed before.

6. William A. Dembski, *The Design Revolution: Answering the Toughest Questions about Intelligent Design* (Downers Grove, IL: InterVarsity, 2004).

7. I literally could write over a hundred pages, and bore you to death in the process, if I had to cover all of the shelling of Dembski. See the listing in "Attacks on Dembski," at the Discovery.org webpage for *Darwin Strikes Back* for summaries of other key attacks on Dembski and the filter, with ID rejoinders.

8. See especially Woodward, *Doubts about Darwin*, 171–82.

9. Dembski came to a lecture by philosopher Alvin Plantinga, Whig-Clio Hall, Princeton University, October 1990. Dembski was doing postdoctoral work at Princeton at the time.

10. See Dembski, *The Design Revolution*, 82–83.

11. See ibid., 84–86. Dembski points out that a number of suggested (in publications) universal probability bounds were in the range of 1 in 10^{94} up to 1 in 10^{120}. Dembski's figure is the most conservative in the literature.

12. See ibid., 76.

13. See the discussion of this point of applying the filter to the flagellum in "The Flagellum Unspun," by Kenneth Miller in *Debating Design*, ed. Dembski and Ruse.

14. Dembski, *The Design Revolution*, 88.

15. Ibid., 95–96.

16. In ibid., chapter 10, p. 81, Dembski quotes Leslie Orgel as the first to use the term in his book, *The Origins of Life: Molecules and Natural Selection* (New York: Wiley, 1973). He also cites Paul Davies's use of the phrase in his book *The Fifth Miracle*, which I referred to in the previous chapter.

17. This personal attack strategy is typical in anti-ID reviews of Dembski. See Forrest and Gross, *Creationism's Trojan Horse*, 118: "The implication is, of course, that he alone has met the immemorial challenge to logic, mathematics, natural science, metaphysics, and moral philosophy, the challenge that had eluded them all until just now: to establish the truth of life's willed designing by an incomprehensibly intelligent agent outside nature." After a mere listing of Dembski's core arguments, the authors assert, with minimal charity and justification, "That is not the voice of a modest young scholar."

18. Quotes are from ibid., 123, and Perakh, *Unintelligent Design*, 26–28.

19. Perakh, *Unintelligent Design*, 104.

20. See Dembski's discussion in *The Design Revolution*, 93.

21. Perakh, *Unintelligent Design*.

22. I am indebted to the late Donald Mackay for this image, even though I am using it in a different way. Donald Mackay, *The Clockwork Image* (Downers Grove, IL: InterVarsity, 1974).

23. This quote is from Dembski, *The Design Revolution*, 98.

24. See Michael Polanyi, "Life Transcending Physics and Chemistry," *Chemical and Engineering News* (August 21, 1967).

25. Dembski, *The Design Revolution*, 93. In the context, Dembski cites the Ruse source (with no page number) as *Can a Darwinian Be a Christian? The Relationship between Science and Religion* (New York: Cambridge University Press, 2001).

26. Quoted from Dembski, *The Design Revolution*, 93, italics added.

27. Ibid., 99, italics added.

28. Neil deGrasse Tyson, "The Perimeter of Ignorance," *Natural History* (November 2005). Tyson will reappear in the next chapter as an atheologian cheerleader for the Darwinists.

29. The various quotes are all from Dembski, *The Design Revolution*, 90.

30. Shanks, *God, the Devil, and Darwin*, 127, 129 (see 125–29). See also Young and Edis, eds., *Why Intelligent Design Fails*, 91–95.

31. Cornelius Hunter, "Can Science Refute Design? A Book Review of *Why Intelligent Design Fails*," *Origins* no. 58 (June 21, 2005): 37.

32. Dembski, *The Design Revolution*. I would list the following chapters as especially packed with strong interaction with Darwinism: chapter 19, "Information Ex Nihilo"; chapters 25–26, "The Supernatural" and "Embodied and Unembodied Designers"; chapter 30, "The Argument from Ignorance;" and chapter 36, "The Only Game in Town."

33. Ibid., 145.

34. Holmes Rolston III, *Genes, Genesis and God: Values and Their Origins in Natural and Human History* (New York: Cambridge University Press, 1999), cited in Dembski, *The Design Revolution*, 146.

35. All the quotes and information in this and the remaining paragraphs in this chapter are from Dembski, *The Design Revolution*, 146–48.

Chapter 11

1. David Berlinski, "The Deniable Darwin," *Commentary* 101, no. 6 (June 1996).

2. For Berlinski's range of thought and writing, see his writings referenced on the Discovery.org website.

3. The other four were paleontologist David Raup, who assisted greatly in my research on Phillip Johnson; historian of science Ron Numbers, who reviewed the manuscript and urged me to get it published through a trade publisher; and two of the four members of my dissertation committee who played a pivotal role and said they enjoyed my work, without siding with ID in any ideological sense.

4. A key example of a theologian siding with evolution and against ID would be John Haught, a theologian at Georgetown University. John Haught wrote a fascinating chapter, "Darwin, Design, and the Divine Providence," in *Debating Design*, ed. Dembski and Ruse. Haught's best-known recent book is *God After Darwin: A Theology of Evolution* (Boulder, CO: Westview Press, 2000), which Michael Behe reviewed in a relatively positive and appreciative tone. See: www.arn.org/docs/behe/mb_godafter darwinreview.htm.

5. The integrity of design arguments depends, to some extent, on accuracy of facts. At least on the level of scientific accuracy, many scientific correspondents of design theorists have helped enormously as fact-checkers. Others have gone far beyond fact-checking and have interacted extensively with Dembski and other theorists as they have willingly critiqued their arguments, which led to progressively greater clarity and strength of those arguments. Virtually all of these friendly critics remain, for obvious reasons, very confidential.

6. Woodward, *Doubts about Darwin*, 73–74, 82–83, 260 n. 46. David Raup is also mentioned in chapter 3 of this book.

7. Two of the Berkeley professors in attendance maintained a cordial and lively correspondence with Johnson afterward; they were civil participants in proto-ID discussions if not "helpers" per se.

8. Most notable in this genre were Niall Shanks, *God, the Devil, and Darwin*, and Barbara Forrest and Paul Gross, *Creationism's Trojan Horse*. Many of Shanks's distortions are dissected in Del Ratsch's penetrating critique "How Not to Critique Intelligent Design Theory," in the web magazine *Ars Disputandi*, vol. 5 (2005). See the appendix for a key excerpt from this review.

9. Here are three reasons for minimal mention of physics and cosmology in *Doubts about Darwin*: First, I faced some tight space limitations and had to cover the main figures of ID adequately before moving on to secondary figures. Second, there always has been a heavy emphasis in ID on *biology*—on the empirical weaknesses in evolutionary biology and on origin-of-life theory. My history reflected that tilt. A third reason is that the earliest books dealing with the discovery of the universe's fine-tuning were not published until the 1980s, so this side of cosmological design was just taking its baby steps at the time ID was birthed in the 1980s.

10. In Young and Edis, eds., *Why Intelligent Design Fails*, physicist Victor Stenger tackles the topic in his chapter, "Is the Universe Fine-Tuned for Us?"

11. These basic facts are drawn from Robert Jastrow, *God and the Astronomers*, 2nd ed. (New York: W. W. Norton, 1992), chapter 2, 17–25.

12. Any of the writings on the big bang by Hugh Ross, president of Reasons to Believe, will sound this tone. This is true whether it be a book, Internet article, or newsletter.

13. From Simon Singh, *Big Bang: The Origin of the Universe* (New York: Fourth Estate, 2004), 483.

14. According to Singh, this expansion of the inflationary epoch was completed within the first 10^{-35} second—that is, the first hundred millionth of a billionth of a billionth of a billionth of a second. Ibid., 477–78 n. 8.

15. Jastrow, *God and the Astronomers*, 2nd ed., 106.

16. See Behe, *Darwin's Black Box*, 244–45. Behe has also made this same point in countless other settings.

17. Jastrow, *God and the Astronomers*, 2nd ed., 107.

18. Ibid., 9.

19. A theoretical openness to theism is suggested by Jastrow's supportive blurb on the back of *The Mystery of Life's Origin*, by Thaxton, Bradley, and Olsen, mentioned in chapter 8.

20. John Barrow and Frank Tipler, *The Anthropic Cosmological Principle* (New York: Oxford University Press, 1986). William Lane Craig, "Barrow and Tipler on the Anthropic Principle vs. Divine Design," published at Craig's Leaderu.com virtual office: www.leaderu.com/offices/billcraig/docs/barrow.html, accessed January 25, 2006.

21. Gonzalez and Richards, *The Privileged Planet*.

22. Astrophysicist Marc Davis is the source for this quote, in his video interview in *Evidence for God*, a documentary published by Fred Heeren, DayStar Publications, 1997.

23. Based on several interviews with Hugh Ross, the astrophysicist who founded Reasons to Believe. These were in early June and late July 2005.

24. The FTU is explored by philosopher John Leslie in *Universes* (New York: Routledge, 1989).

25. This is from the Wikipedia page on "Anthropic principle" as accessed in January 2006. (When the page was accessed again in June 2006, the wording had been changed somewhat.) The 1988 reference is to Hawking's book *A Brief History of Time* (NY: Bantam Books).

26. Timothy Ferris, *The Whole Shebang: A State-of-the-Universe(s) Report* (New York: Simon and Schuster, 1997).

27. The phrase "bloated ontology" is from William Dembski, in a November 2000 email summarizing his chat with a Yale cosmologist who attended the Yale Design Conference. The professor attended the talk by William Lane Craig on design in the universe and defended the Multiverse afterward in discussion with ID theorists.

28. Simon Conway-Morris, quoted by Douglas A. Vakoch, in *Nature* 429, no. 6994 (24 June 2004): 808.

29. Richard Dawkins, in Shanks, *God, the Devil, and Darwin*, viii-ix.

30. See Woodward, *Doubts about Darwin*, 136–41, where I discuss David Hull's evil-and-waste argument.

31. Cornelius Hunter, *Darwin's God: Evolution and the Problem of Evil* (Grand Rapids: Brazos Press, 2001).

32. See, for example, the chapter "On the Design of the Vertebrate Retina," by George Ayoub, in *Darwinism Under the Microscope*, ed. Thomas Woodward and James Gills (Lake Mary, FL: Strang, 2002).

33. Dembski, *The Design Revolution*, 58.

34. Ibid.

35. Neil deGrasse Tyson, "The Perimeter of Ignorance," *Natural History* (November 2005). Among celestial dangers, besides matter-hungry black holes and colliding galaxies, his nightmare features the "shooting gallery, full of rogue asteroids and comets that collide with planets from time to time."

36. Ibid.

37. Ibid.

Chapter 12

1. See Niall Shanks, *God, the Devil, and Darwin*, and Barbara Forrest and Paul Gross, *Creationism's Trojan Horse*. Both were released in 2004 by Oxford University Press.

2. See Charles Krauthammer's column "Phony Theory, False Conflict" in the November 18, 2005 *Washington Post*. Also, see George Will's column "Grand Old Spenders" in the same paper, on the previous day (November 17, 2005), as well as his column "Evolution Debate Will Not End" in the July 4, 2005 issue of *Newsweek*.

3. This *ad hominem* (against the man) attack is related to the genetic fallacy, where one faults a theory by some negative aspect of how one learned or developed a belief. Both are related, in turn, to "poisoning the well"—as seen on Wikipedia.org/wiki/Poisoning_the_well: "Poisoning the well is a logical fallacy where adverse information about someone is preemptively presented to an audience, with the intention of discrediting or ridiculing everything that person is about to say. Poisoning the well is a special case of *argumentum ad hominem*."

4. In chapter 10 of Woodward, *Doubts about Darwin*, "recalcitrant nature" is the ultimate power player in the drama of origins.

5. Franklin Harold, *The Way of the Cell: Molecules, Organisms, and the Order of Life* (New York: Oxford University Press, 2001).

6. See Meyer, "The Origin of Biological Information and the Higher Taxonomic Categories," 217. Meyer references Gerhart and Kirschner, "Cells, Embryos, and Evolution," *Blackwell Science* (London, 1997): 121, and M. D. Adams, M.D., "The Genome Sequence of Drosophila Melanogaster," *Science* 287 (2000): 2185–95.

7. To review his talk, see Ralph Seelke, "What Can Evolution Really Do? How Microbes Can Help Us Find the Answer," in the Uncommon Dissent Forum, August 2005, Greenville, SC. Copies are available through Lewis Young at Piedmonttravel.com. Seelke's work at the University of Wisconsin–Superior refutes the charge that ID scholars don't do experimental research. See also Doug Axe's article, "Estimating the Prevalence of Protein Sequences Adopting Functional Enzyme Folds," *Journal of Molecular Biology* 341, no. 5, 1295–1315.

8. Muller and Newman, eds., *Origination of Organismal Form*, 7.

9. Philip Skell, "Why Do We Invoke Darwin?," *The Scientist*, August 29, 2005, p. 10.

10. The information in this and the preceding paragraph is taken from Michael Behe, "Whether Intelligent Design Is Science," February 3, 2006, www.discovery.org. This was later incorporated as an appendix in *Traipsing into Evolution*.

11. Darwin, *Origin of the Species*, 4, italics added.

Index

Thomas Woodward (Ph.D., University of South Florida) is a professor at Trinity College of Florida, where he teaches the history of science, philosophy, communication, and systematic theology. He is founder and director of the C. S. Lewis Society and lectures in universities on scientific, apologetic, and religious topics. The author of the award-winning *Doubts about Darwin*, Woodward is an avid astronomer and has been published in *Christianity Today* and other periodicals. He lives in Dunedin, Florida.

Also available *from* Thomas Woodward

A *Christianity Today* Book Award Winner

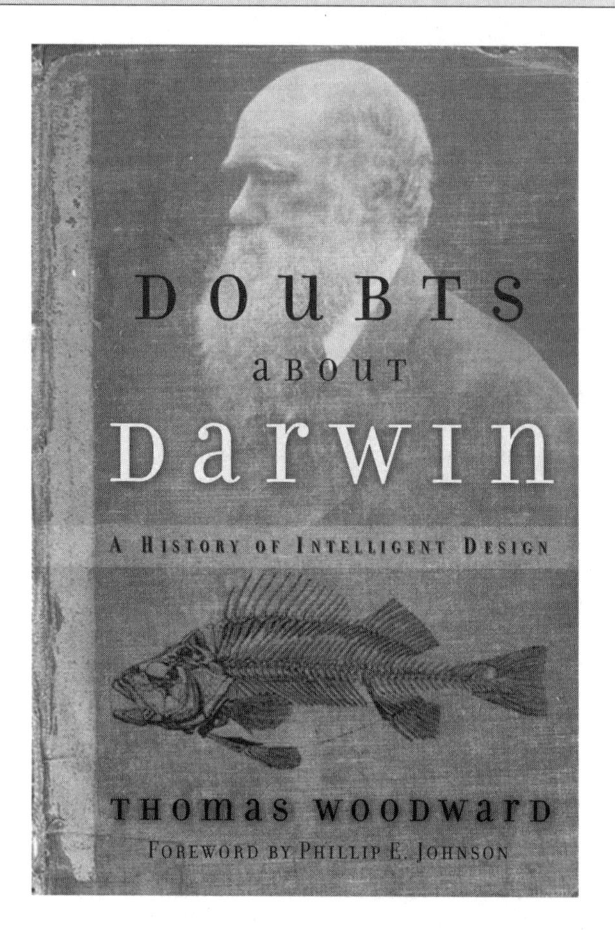

Deep in the halls of scientific academia, a debate has been quietly raging between Darwinian macroevolution and the theory of Intelligent Design. *Doubts about Darwin* follows the key players in this debate, confronts the challenge of scientific naturalism, and provides a stirring overview of this growing movement.

"*Doubts about Darwin* is an exciting history lesson. While there are no truces in view, these fighters are working toward intellectual freedom. And their stories can inspire you as you face your school board, colleagues, or biology professors."

—Charles Colson, BreakPoint

BakerBooks

Available at yo ... om